A CONCISE GEOMETRY

BY

CLEMENT V. DURELL, M.A.

SENIOR MATHEMATICAL MASTER, WINCHESTER COLLEGE

LONDON
G. BELL AND SONS LTD.
1921

First Published . . January 1921
Reprinted . . . . . August 1921

Printing Statement:

Due to the very old age and scarcity of this book, many of the pages may be hard to read due to the blurring of the original text, possible missing pages, missing text, dark backgrounds and other issues beyond our control.

Because this is such an important and rare work, we believe it is best to reproduce this book regardless of its original condition.

Thank you for your understanding.

PREFACE

THE primary object of this text-book is to supply a large number of easy examples, numerical and theoretical, and as varied in character as possible, in the belief that the educational value of the subject lies far more in the power to apply the fundamental facts of geometry, and reason from them, than in the ability to reproduce proofs of these facts. This collection has grown out of a set of privately printed geometrical exercises which has been in use for many years at Winchester College: the author is indebted to many friends for additions to it, and to the following authorities for permission to use questions taken from examination papers:—
The Controller of His Majesty's Stationery Office; The Syndics of the Cambridge University Press; and the Oxford and Cambridge Joint Board.

Riders are arranged in exercises corresponding to groups of theorems, and Constructions are treated similarly. There is also a set of fifty Revision Papers. Answers are given to all numerical questions, except where no intermediate work is necessary. Harder questions and papers are marked with an asterisk.

The book is arranged as follows:—
    I. Riders, Numerical and Theoretical.
    II. Practical Geometry; Construction Exercises.
    III. Proofs of Theorems.
    IV. Proofs of Constructions.

The Proofs of Theorems and Constructions are collected together instead of being dispersed through the book in order to assist revision by arranging the subject-matter in a compact form. When learning or revising proofs of theorems and constructions, it is most important the student should draw his own rough figure. For this reason, no attempt has been made to arrange the whole of the proof of a theorem on the same page as the figure corresponding to it.

The order and method of proof is arranged to suit those who are revising for examination purposes, and is not intended to be that used in a first course. It is now generally agreed that proofs by superposition of congruence tests and proofs of the fundamental angle property of parallel lines should be omitted in the preliminary course, but that these facts should be assumed without formal proof and utilised for simple applications, the former being treated by some such method as that noted on page 14, and the latter by rotation or the set-square method of drawing parallel lines. This broadens the basis of the geometrical work and enables the early exercises to be of a more interesting nature.

The arrangement of riders in one group and practical work in another is made for convenience of reference. Naturally both groups will be in use simultaneously; but the course should open with the exercise on the use of instruments in the practical geometry section.

No attempt has been made to include in the text the usual preliminary oral instruction which deals with the fundamental concepts of angles, lines, planes, surfaces, solids, and requires illustration with simple models. The examples start with methods of measurement and general use of instruments, which is the earliest stage at which a book is really any use for class work. The object throughout has been to arrange the book to suit the student rather than the teacher, and "talk" is therefore cut down to the minimum. It is the nature of the examples which has been the chief consideration, and if this part of the book receives approval, the author will consider his object has been attained.

Valuable assistance has been given by Mr. A. E. BROOMFIELD, without whose advice, interest, and encouragement the work could scarcely have been carried out.

C. V. D.

August 1920.

CONTENTS

RIDERS ON BOOK I

	PAGE
ANGLES AT A POINT	1
ANGLES AND PARALLEL LINES	5
ANGLES OF A TRIANGLE, ETC.	9
ISOSCELES TRIANGLES, CONGRUENT TRIANGLES (FIRST SECTION)	15
CONGRUENT TRIANGLES (SECOND SECTION), ETC.	21

RIDERS ON BOOK II

AREAS	25
PYTHAGORAS' THEOREM	37
EXTENSION OF PYTHAGORAS' THEOREM	43
SEGMENTS OF A STRAIGHT LINE	46
INEQUALITIES	48
INTERCEPT THEOREM	51

RIDERS ON BOOK III

SYMMETRICAL PROPERTIES OF A CIRCLE	56
ANGLE PROPERTIES OF A CIRCLE	60
ANGLE PROPERTIES OF TANGENTS	67
PROPERTIES OF EQUAL ARCS	71
LENGTHS OF TANGENTS, CONTACT OF CIRCLES	76
CONVERSE PROPERTIES	82
MENSURATION	86
LOCI	93
CIRCUMCIRCLE	99
IN-CIRCLE, EX-CIRCLES	100
ORTHOCENTRE	101
CENTROID	103

RIDERS ON BOOK IV

PROPORTION	105
SIMILAR TRIANGLES	111

CONTENTS

	PAGE
RECTANGLE PROPERTIES OF A CIRCLE	120
AREAS AND VOLUMES (SIMILAR FIGURES, SOLIDS)	126
THE BISECTOR OF AN ANGLE OF A TRIANGLE	131

CONSTRUCTION EXERCISES—BOOK I

USE OF INSTRUMENTS	135
DRAWING TO SCALE	145
MISCELLANEOUS—I	148
TRIANGLES, PARALLELOGRAMS, ETC.	150
MISCELLANEOUS—II	153

CONSTRUCTION EXERCISES—BOOK II

AREAS	155
SUBDIVISION OF A LINE	158

CONSTRUCTION EXERCISES—BOOK III

CIRCLES	161
MISCELLANEOUS—III	172

CONSTRUCTION EXERCISES—BOOK IV

PROPORTION, SIMILAR FIGURES	175
MEAN PROPORTIONAL	178
MISCELLANEOUS—IV	180
REVISION PAPERS, I–L	181

PROOFS OF THEOREMS

BOOK I	205
BOOK II	218
BOOK III	234
BOOK IV	254

PROOFS OF CONSTRUCTION

BOOK I	267
BOOK II	276
BOOK III	280
BOOK IV	289
NOTES	305
GLOSSARY AND INDEX	313
ANSWERS	315

A CONCISE GEOMETRY

RIDERS ON BOOK I

ANGLES AT A POINT

Theorem 1

(i) If a straight line **CE** cuts a straight line **ACB** at **C**, then ∠ **ACE** + ∠ **BCE** = 180°.
(ii) If lines **CA, CB** are drawn on opposite sides of a line **CE** such that ∠ **ACE** + ∠ **BCE** = 180°, then **ACB** is a straight line.

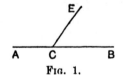

Fig. 1.

Theorem 2

If two straight lines intersect, the vertically opposite angles are equal.

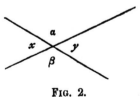

Fig. 2.

$x = y$ and $a = \beta$.

ANGLES AT A POINT

EXERCISE I

1. What are the supplements of 20°, 150°, 27° 45′, 92° 10′ ?
2. What are the complements of 75°, 30° 30′, 10° 48′ ?
3. A wheel has six spokes, what is the angle between two adjacent spokes ?
4. Guess the sizes of the following angles :—

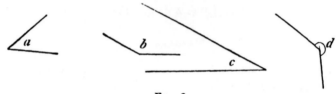

FIG. 3.

5. What is the least number of times you must turn through 17° in order to turn through (i) an obtuse angle, (ii) a reflex angle, (iii) more than one complete revolution ?
6. What is the angle between N.E. and S.E. ?
7. What is the angle between S.S.W. and E.N.E. ?
8. What is the angle between (i) 12° N. of W. and 5° E. of N. ; (ii) S.W. and E.S.E. ; (iii) 22° S. of W. and 9° N. of E. ?
9. Through what angle does the minute hand of a clock turn in 15 minutes, 5 minutes, 20 minutes, 50 minutes, 2 hours 45 minutes ?
10. Through what angle does the hour hand of a clock turn in 40 minutes, 1 hour ?
11. Through what angle has the hour hand of a clock turned, when the minute hand has turned through 30° ?
12. What is the angle between the hands of a clock at (i) 4 o'clock, (ii) ten minutes past four ?
13. A wheel makes 20 revolutions a minute, through what angle does a spoke turn each second ?
14. What equation connects x and y if $x°$ and $y°$ are (i) complementary, (ii) supplementary ?

ANGLES AT A POINT

15. The line **OA** cuts the line **BOC** at **O**; if ∠ **AOB** = 2 ∠ **AOC**, calculate ∠ **AOB**.
16. What angle is equal to four times its complement?
17. A man watching a revolving searchlight notes that he is in the dark four times as long as in the light, what angle of country does the searchlight cover at any moment?
18. The weight in a pendulum clock falls 4 feet in 8 days; through what angle does the hour hand turn when the weight falls 1 inch?
19. What is the reflex angle between the directions S.W. and N.N.W.?
20. If the earth makes one complete revolution every 24 hours, through what angle does it turn in 20 minutes?
21. The longitude of Boston is 71° W., and of Bombay is 73° E., what is their difference of longitude?
22. The latitude of Sydney is 33° S., and of New York is 41° N., what is their difference of latitude?
23. Cape Town has latitude 33° 40′ S. and longitude 18° 30′ E., Cologne has latitude 50° 55′ N. and longitude 7° E., what is their difference of latitude and longitude?
24. ∠ **POQ** = $2x°$, ∠ **QOR** = $3x°$, ∠ **POR** = $4x°$; find x.

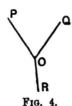

Fig. 4.

25. **OP, OQ, OR, OS** are 4 lines in order such that ∠ **POQ** = 68°, ∠ **QOR** = 53°, ∠ **ROS** = 129°; find ∠ **SOP**. Find also the angle between the lines bisecting ∠ **POS**, ∠ **QOR**.
26. **OA, OB, OC** are 3 lines in order such that ∠ **AOB** = 54°, ∠ **BOC** = 24°; **OP** bisects ∠ **AOC**; find ∠ **POB**.
27. **CD** is perpendicular to **ACB**; **CE** is drawn so that ∠ **DCE** = 23°; find the difference between ∠ **ACE** and ∠ **BCE**. What is their sum?

28. Given $\angle AOD = 145°$, $\angle BOC = 77°$, and $\angle AOB = \angle COD$; calculate $\angle AOC$ (Fig. 5).

Fig. 5.

29. OA, OB, OC, OD, OE, OF are 6 lines in order such that $\angle AOB = 43°$, $\angle BOC = 67°$, $\angle COD = 70°$, $\angle DOE = 59°$, $\angle EOF = 51°$; prove that AOD and COF are straight lines. Calculate the angle between the lines bisecting $\angle AOF$ and $\angle BOC$.

30. $\angle AOB = 38°$; AO is produced to C; OP bisects $\angle BOC$; calculate reflex angle AOP.

31. OA, OB, OC, OD are 4 lines in order such that $\angle AOC = 90° = \angle BOD$; if $\angle BOC = x°$, calculate $\angle AOD$.

32. Two lines AOB, COD intersect at O; OP bisects $\angle AOC$; if $\angle BOC = x°$, calculate $\angle DOP$.

33. OA, OC make acute angles with OB on opposite sides; OP bisects $\angle BOC$; prove $\angle AOB + \angle AOC = 2\angle AOP$.

34. The line OA cuts the line BOC at O; OP bisects $\angle AOB$; OQ bisects $\angle AOC$; prove $\angle POQ = 90°$.

35. OA, OB, OC, OD are 4 lines in order such that $\angle AOB = \angle COD$ and $\angle BOC = \angle AOD$; prove that AOC is a straight line.

36. Given $\angle AOB = \angle DOC$, and that OP bisects $\angle AOD$ (see Fig. 6), if PO is produced to Q, prove that QO bisects $\angle BOC$.

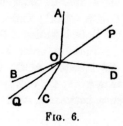

Fig. 6.

ANGLES AND PARALLEL LINES

Theorem 5

In Fig. 7,
(i) If ∠ PBC = ∠ BCS, then PQ is parallel to RS.
ii) If ∠ ABQ = ∠ BCS, then PQ is parallel to RS.
ii) If ∠ QBC + ∠ BCS = 180°, then PQ is parallel to RS.

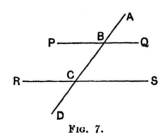

Fig. 7.

Theorem 6

In Fig. 7,
If PQ is parallel to RS,
Then (i) ∠ PBC = ∠ BCS (alternate angles).
(ii) ∠ ABQ = ∠ BCS (corresponding angles).
(iii) ∠ QBC + ∠ BCS = 180°.

CONCISE GEOMETRY

ANGLES AND PARALLEL LINES

EXERCISE II

1. In the following figures, a line cuts two parallel lines. What equations connect the marked angles? Give reasons.

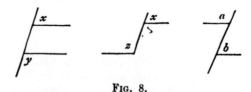

Fig. 8.

2. The following figures contain pairs of parallel lines. What equations connect the marked angles? Give reasons.

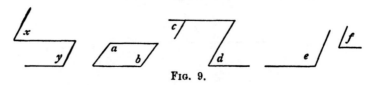

Fig. 9.

3. (i) If one angle of a parallelogram is 60°, find its other angles.
   (ii) If one angle of a parallelogram is 90°, find its other angles.

4. If **AB** is parallel to **ED**, see Fig. 10, prove that ∠ BCD = ∠ ABC + ∠ CDE.

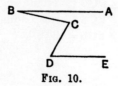

Fig. 10.

5. The side **AB** of the triangle **ABC** is produced to **D**; **BX** is drawn parallel to **AC**; ∠ BAC = 32°, ∠ BCA = 47°; find the remaining angles in the figure and the value of ∠ BAC + ∠ BCA + ∠ ABC.

ANGLES AND PARALLEL LINES

6. If **AB** is parallel to **DE**, see Fig. 11, prove that ∠ **ABC** + ∠ **CDE** = 180° + ∠ **BCD**.

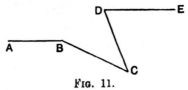

Fig. 11.

[Draw **CF** parallel to **DE**.]

7. In Fig. 12, prove that **AB** is parallel to **ED**.

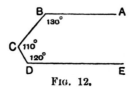

Fig. 12.

8. In Fig. 13, if ∠ **ABC** = 74°, ∠ **EDC** = 38°, ∠ **BCD** = 36°, prove **ED** is parallel to **AB**.

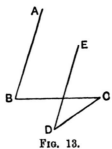

Fig. 13.

9. **ABCD** is a quadrilateral; if **AB** is parallel to **DC**, prove that ∠ **DAB** − ∠ **DCB** = ∠ **ABC** − ∠ **ADC**.

10. In Fig. 14, if **AB** is parallel to **DE**, prove that $x + y - z$ equals two right angles.

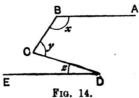

Fig. 14.

11. A line **AC** cuts two parallel lines **AB, CD**; **B** and **D** are on the same side of **AC**; the lines bisecting the angles **CAB, ACD** meet at **O**; prove that $\angle \mathbf{AOC} = 90°$.
12. If two straight lines are each parallel to the same straight line, prove that they are parallel to each other.

ANGLES OF A TRIANGLE AND OTHER RECTILINEAL FIGURES

THEOREM 7

(i) If the side **BC** of the triangle **ABC** is produced to **D**, $\angle ACD = \angle BAC + \angle ABC$.

(ii) In the $\triangle ABC$, $\angle ABC + \angle BCA + \angle CAB = 180°$.

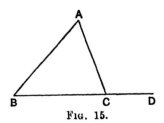

FIG. 15.

THEOREM 8

(i) The sum of the interior angles of any convex polygon which has n sides is $2n - 4$ right angles.

(ii) If the sides of a convex polygon are produced in order, the sum of the exterior angles is 4 right angles.

ANGLES OF A TRIANGLE AND OTHER RECTILINEAL FIGURES

EXERCISE III

1. In a right-angled triangle, one angle is 37°, what is the third angle?
2. Two angles of a triangle are each 53°, what is the third angle?
3. If ∠BAC = 43° and ∠ABC = 109°, what is ∠ACB?
4. The side BC of the triangle ABC is produced to D; ∠ABD = 19°, ∠ACD = 37°, what is ∠BAC?
5. In the quad. ABCD, ∠ABC = 112°, ∠BCD = 75°, ∠DAB = 51°, what is ∠CDA?
6. ABCD is a straight line and P a point outside it; ∠PBA = 110°, ∠PCD = 163°, find ∠BPC.
7. Three of the angles of a quad. are equal; the fourth angle is 120°; find the others.
8. Can a triangle be drawn having its angles equal to (i) 43°, 64°, 73°; (ii) 45°, 65°, 80°?
9. What is the remaining angle of a triangle, if two of its angles are (i) 120°, 40°; (ii) 50°, x°; (iii) $2x$°, $3x$°; (iv) $x+10$, $20-x$ degrees?
10. The angles of a triangle are x°, $2x$°, $2x$°; find x.
11. If in the triangle ABC, ∠BAC = ∠BCA + ∠ABC, find ∠BAC.
12. If A, B, C are the angles of a triangle and if $A - B = 15°$, $B - C = 30°$, find A.
13. The angles of a five-sided figure are x, $2x$, $x+30$, $x-10$, $x+40$ degrees, find x.
14. The angles of a pentagon are in the ratio 1 : 2 : 3 : 4 : 5; find them.
15. In △ABC, ∠ABC = 38°, ∠ACB = 54°; AD is perpendicular to BC; AE bisects ∠BAC, find ∠EAD.
16. In △ABC, ∠BAC = 74°, ∠ABC = 28°; BC is produced to X; the lines bisecting ∠ABC and ∠ACX meet at K. Find ∠BKC.

ANGLES OF A TRIANGLE

17. In △ABC, ∠ABC = 32°, ∠BAC = 40°; find the angle at which the bisector of the greatest angle of the triangle cuts the opposite side.

18. In △ABC, ∠ABC = 110°, ∠ACB = 50°; AD is the perpendicular from A to CB produced; prove that ∠DAB = ∠BAC.

19. The base BC of △ABC is produced to D; if ∠ABC = ∠ACB and if ∠ACD = $x°$, calculate ∠BAC.

20. In the quad. ABCD, ∠ABC = 140°, ∠ADC = 20°; the lines bisecting the angles BAD, BCD meet at O; calculate ∠AOC.

21. In △ABC, the bisector of ∠BAC cuts BC at D, if ∠ABC = $x°$ and ∠ACB = $y°$, calculate ∠ADC.

22. If the angles of a quad. taken in order are in the ratio 1 : 3 : 5 : 7, prove that two of its sides are parallel.

23. Each angle of a polygon is 140°; how many sides has it?

24. Find the sum of the interior angles of a 12-sided convex polygon.

25. Find the interior angle of a regular 20-sided figure.

26. Prove that the sum of the interior angles of an 8-sided convex polygon is twice the sum of those of a pentagon.

27. Each angle of a regular polygon of x sides is $\frac{3}{4}$ of each angle of a regular polygon of y sides; express y in terms of x, and find any values of x, y which will fit.

28. The sum of the interior angles of an n-sided convex polygon is double the sum of the exterior angles. Find n.

29. In Fig. 16, prove that $x = a + b - y$.

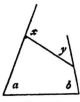

Fig. 16.

30. In Fig. 17, prove that $x - y = a - b$.

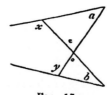

Fig. 17.

31. In Fig. 18, express x in terms of a, b, c.

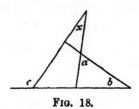

Fig. 18.

32. In Fig. 19, express x in terms of a, b, c.

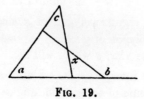

Fig. 19.

33. In Fig. 20, express x in terms of a, b, c.

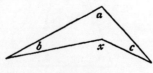

Fig. 20.

ANGLES OF A TRIANGLE

34. If, in Fig. 21, $x+y=3z$, prove that the triangle is right-angled.

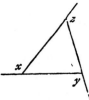

Fig. 21.

35. Prove that the reflex angles in Fig. 22 are connected by the relation $a+b=x+y$.

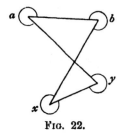

Fig. 22.

36. D is a point on the base BC of the triangle ABC such that ∠ DAC = ∠ ABC, prove that ∠ ADC = ∠ BAC.
37. The diagonals of the parallelogram ABCD meet at O, prove that ∠ AOB = ∠ ADB + ∠ ACB.
38. If, in the quadrilateral ABCD, AC bisects the angle DAB and the angle DCB, prove that ∠ ADC = ∠ ABC.
39. ABC is a triangle, right-angled at A; AD is drawn perpendicular to BC, prove that ∠ DAC = ∠ ABC.
40. ABCD is a parallelogram, prove that the lines bisecting the angles DAB, DCB are parallel.
41. In the △ABC, BE and CF are perpendiculars from B, C to AC, AB; BE cuts CF at H; prove that ∠ CHE = ∠ BAC.
42. If, in the quadrilateral ABCD, ∠ ABC = ∠ ADC and ∠ BCD = ∠ BAD, prove that ABCD is a parallelogram.
43. If in the △ABC the bisectors of the angles ABC, ACB meet at I, prove that ∠ BIC = $90° + \frac{1}{2}$ ∠ BAC.

44. The side **BC** of the triangle **ABC** is produced to **D**; **CP** is drawn bisecting ∠ **ACD**; if ∠ **CAB** = ∠ **CBA**, prove that **CP** is parallel to **AB**.

45. The side **BC** of △**ABC** is produced to **D**; the lines bisecting ∠ **ABC**, ∠ **ACD** meet at **Q**; prove that ∠ **BQC** = $\frac{1}{2}$ ∠ **BAC**.

46. The base **BC** of △**ABC** is produced to **D**; the bisector of ∠ **BAC** cuts **BC** at **K**; prove ∠ **ABD** + ∠ **ACD** = 2 ∠ **AKD**.

47. The sides **AB**, **AC** of the triangle **ABC** are produced to **H**, **K**; the lines bisecting ∠ **HBC**, ∠ **KCB** meet at **P**; prove that ∠ **BPC** = 90° − $\frac{1}{2}$ ∠ **BAC**.

48. **P** is any point inside the triangle **ABC**, prove that ∠ **BPC** > ∠ **BAC**.

49. In the quadrilateral **ABCD**, the lines bisecting ∠ **ABC**, ∠ **BCD** meet at **P**, prove that ∠ **BAD** + ∠ **CDA** = 2 ∠ **BPC**.

CONGRUENT TRIANGLES

Given a triangle **ABC**, what set of measurements must be made in order to copy it?

1. Measure **AB**, **AC**, ∠ **BAC**.
    This is enough to fix the size and shape of the triangle. Therefore all triangles drawn to these measurements will be congruent to △ **ABC** and to each other.
    This result is given as Theorem 3.
2. Measure **BC**, ∠ **ABC**, ∠ **ACB**.
    This also fixes the triangle. [Theorem 9.]
3. Measure **BC**, **CA**, **AB**.
    This also fixes the triangle. [Theorem 11.]

ISOSCELES TRIANGLES AND CONGRUENT TRIANGLES (First Section)

Theorem 3

In the triangles ABC, PQR,
 If AB = PQ, AC = PR, ∠ BAC = ∠ QPR,
  Then △ABC ≡ △PQR.

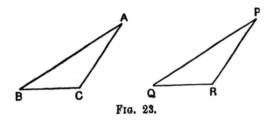

Fig. 23.

Theorem 9

In the triangles ABC, PQR,
 (i) If BC = QR, ∠ ABC = ∠ PQR, ∠ ACB = ∠ PRQ,
  Then △ABC ≡ △PQR.
 (ii) If BC = QR, ∠ ABC = ∠ PQR, ∠ BAC = ∠ QPR,
  Then △ABC ≡ △PQR.

Theorem 10

ABC is a triangle.
 (i) If AB = AC, then ∠ ACB = ∠ ABC
 (ii) If ∠ ACB = ∠ ABC, then AB = AC.

Fig. 24.

ISOSCELES TRIANGLES AND CONGRUENT TRIANGLES (First Section)

EXERCISE IV

1. The vertical angle of an isosceles triangle is 110°; what are the base angles?
2. One base angle of an isosceles triangle is 62°; what is the vertical angle?
3. Find the angles of an isosceles triangle if (i) the vertical angle is double a base angle, (ii) a base angle is double the vertical angle.
4. In the triangle ABC, ∠ BAC = 2 ∠ ABC and ∠ ACB − ∠ ABC = 36°; prove that the triangle is isosceles.
5. A, B, C are three points on a circle, centre O; ∠ AOB = 100°, ∠ BOC = 140°, calculate the angles of the triangle ABC.
6. In Fig. 25, if AB = AC, find x in terms of y.

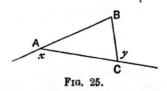

Fig. 25.

7. D is a point on the base BC of the isosceles triangle ABC such that BD = BA; if ∠ BAD = $x°$ and ∠ DAC = $y°$, express x in terms of y.
8. ABCDE is a regular pentagon, prove that the line bisecting the angle BAC is perpendicular to AE.
9. In the triangle ABC, AB = AC; D is a point in AC such that AD = BD = BC. Calculate ∠ BAC.
10. If the line PQ bisects AB at right angles, prove that PA = PB.
11. Two unequal lines AC, BD bisect each other, prove that AB = CD.
12. In the quadrilateral ABCD, AB is equal and parallel to DC; prove that AD is equal and parallel to BC.
13. A line AP is drawn bisecting the angle BAC; PX, PY are the perpendiculars from P to AB, AC; prove that PX = PY.

ISOSCELES TRIANGLES

14. D is the mid-point of the base BC of the triangle ABC, prove that B and C are equidistant from the line AD.

15. A straight line cuts two parallel lines at A, B; C is the mid-point of AB; any line is drawn through C cutting the parallel lines at P, Q; prove that PC = CQ.

16. If the diagonal AC of the quadrilateral ABCD bisects the angles DAB, DCB, prove that AC bisects BD at right angles.

17. ABCD is a quadrilateral; E, F are the mid-points of AB, CD; if $\angle$ AEF = 90° = $\angle$ EFD, prove that AD = BC.

18. The diagonals of a quadrilateral bisect each other at right angles, prove that all its sides are equal.

19. Two lines POQ, ROS bisect each other, prove that the triangles PRS, PQS are equal in area.

20. Two lines POQ, ROS intersect at O; SP and QR are produced to meet at T; if OP = OR and OS = OQ, prove TS = TQ.

21. ABC is an equilateral triangle; BC is produced to D so that BC = CD; prove that $\angle$ BAD = 90°.

22. In the $\triangle$ ABC, AB = AC; AB is produced to D so that BD = BC; prove that $\angle$ ACD = 3 ADC.

23. P is a point on the line bisecting $\angle$ BAC; through P, a line is drawn parallel to AC and cutting AB at Q; prove AQ = QP.

24. In $\triangle$ ABC, AB = AC; D is a point on AC produced such that BD = BA; if $\angle$ CBD = 36°, prove BC = CD.

25. If in Fig. 26, AB = AC and CP = CQ, prove $\angle$ SRP = 3 $\angle$ RPC.

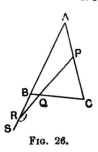

Fig. 26.

26. The base BC of the isosceles triangle ABC is produced to D; the lines bisecting $\angle$ ABC and $\angle$ ACB meet at I; prove $\angle$ ACD = $\angle$ BIC.

27. In the quadrilateral ABCD, AB = AD and ∠ABC = ∠ADC, prove CB = CD.

28. ABC is an acute-angled triangle; AB < AC; the circle, centre A, radius AB cuts BC at D, prove that ∠ABC + ∠ADC = 180°.

29. A, B, C are three points on a circle, centre O; prove ∠ABC = ∠OAB + ∠OCB.

30. AB, AC are two chords of a circle, centre O; if ∠BAC = 90°, prove that BOC is a straight line.

31. In the △ABC, AB = AC; the bisectors of the angles ABC and ACB meet at I, prove that IB = IC.

32. AD is an altitude of the equilateral triangle ABC; ADX is another equilateral triangle, prove that DX is perpendicular to AB or AC.

33. BC is the base of an isosceles triangle ABC; P, Q are points on AB, AC such that AP = PQ = QB = BC; calculate ∠BAC.

34. D is the mid-point of the base BC of the triangle ABC; if AD = DB, prove ∠BAC = 90°.

35. In the quadrilateral ABCD, AB = CD and ∠ABC = ∠DCB, prove ∠BAD = ∠CDA.

36. In the △ABC, AB > AC; D is a point on AB such that AD = AC; prove ∠ABC + ∠ACB = 2∠ADC.

37. The triangle ABC is right-angled at A; AD is the perpendicular from A to BC; P is a point on CB such that CP = CA; prove AP bisects ∠BAD.

38. The vertical angles of two isosceles triangles are supplementary; prove that their base angles are complementary.

39. Draw two triangles ABC, XYZ which are such that AB = XY, AC = XZ, ∠ABC = ∠XYZ but are not congruent. Prove ∠ACB + ∠XZY = 180°.

40. In the △ABC, AB = AC; P is any point on BC produced; PX, PY are the perpendiculars from P to AB, AC produced; prove ∠XPB = ∠YPB.

41. ABC is any triangle; ABX, ACY are equilateral triangles external to ABC; prove CX = BY.

42. OA = OB = OC and ∠BAC is acute; prove ∠BOC = 2∠BAC.

43. In the △ABC, AB = AC; AB is produced to D; prove ∠ACD − ∠ADC = 2∠BCD.

ISOSCELES TRIANGLES

44. D is a point on the side AB of △ABC such that AD = DC = CB; AC is produced to E; prove ∠ECB = 3∠ACD.
45. In the △ABC, ∠BAC is obtuse; the perpendicular bisectors of AB, AC cut BC at X, Y; prove ∠XAY = 2∠BAC − 180°.
46. In the △ABC, AB = AC and ∠BAC > 60°; the perpendicular bisector of AC meets BC at P; prove ∠APB = 2∠ABP.
47. D is the mid-point of the side AB of △ABC; the bisector of ∠ABC cuts the line through D parallel to BC at K; prove ∠BKA = 90°.
48. In the △ABC, ∠BAC = 90° and AB = AC; P, Q are points on AB, AC such that AP = AQ; prove that the perpendicular from A to BQ bisects CP.
49. X, Y are the mid-points of the sides AB, AC of the △ABC; P is any point on a line through A parallel to BC; PX, PY are produced to meet BC at Q, R; prove QR = BC.
50. ABC is a triangle; the perpendicular bisectors of AB, AC meet at O; prove OB = OC.
51. ABC is a triangle; the lines bisecting the angles ABC, ACB meet at I; prove that the perpendiculars from I to AB, AC are equal.
52. The sides AB, AC of the triangle ABC are produced to H, K; the lines bisecting the angles HBC, KCB meet at I; prove that the perpendiculars from I to AH, AK are equal.
53. Two circles have the same centre; a straight line PQRS cuts one circle at P, S and the other at Q, R; prove PQ = RS.
54. ABC is a △; a line AP is drawn on the same side of AC as B, meeting BC at P, such that ∠CAP = ∠ABC; a line AQ is drawn on the same side of AB as C, meeting BC at Q, such that ∠BAQ = ∠ACB; prove AP = AQ.
55. The line joining the mid-points E, F of AB, AC is produced to G so that EF = FG; prove that BE is equal and parallel to CG.
56. In the 5-sided figure ABCDE, the angles at A, B, C, D are each 120°; prove that AB + BC = DE.
57. ABC is a triangle; lines are drawn through C parallel to the bisectors of the angles CAB, CBA to meet AB produced in D, E; prove that DE equals the perimeter of the triangle ABC.

58. **AB, BC, CD** are chords of a circle, centre **O**; if $\angle$ **AOB** $= 108°$ $\angle$ **BOC** $= 60°$, $\angle$ **COD** $= 36°$, prove **AB** $=$ **BC** $+$ **CD**. [From **BA** cut off **BQ** equal to **BO**: join **OQ**.]
59. In the triangles **ABC, XYZ**, if **BC** $=$ **YZ**, $\angle$ **ABC** $= \angle$ **XYZ**, **AB** $+$ **AC** $=$ **XY**, prove $\angle$ **BAC** $= 2 \angle$ **YXZ**.
60. In the $\triangle$ **ABC, AB** $=$ **AC** and $\angle$ **ABC** $= 2 \angle$ **BAC**; **BC** is produced to **D** so that $\angle$ **CAD** $= \frac{1}{2} \angle$ **BAC**; **CF** is the perpendicular from **C** to **AB**; prove **AD** $= 2$**CF**.

CONGRUENT TRIANGLES (Second Section), PARALLELOGRAMS, SQUARES, Etc.

Theorem 11

In the triangles **ABC, XYZ**,
  If **AB = XY, BC = YZ, CA = ZX**,
  Then   △**ABC** ≡ △**XYZ**.

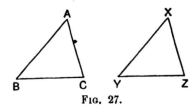

Fig. 27.

Theorem 12

In the triangles **ABC, XYZ**,
  If ∠**ABC** = 90° = ∠**XYZ, AC = XZ, AB = XY**,
  Then   △**ABC** ≡ △**XYZ**.

Theorem 13

If **ABCD** is a parallelogram,
  Then (i)   **AB = CD** and **AD = BC**.
    (ii) ∠**DAB** = ∠**DCB** and ∠**ABC** = ∠**ADC**.
    (iii) **BD** bisects **ABCD**.

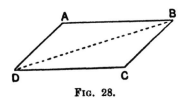

Fig. 28.

Theorem 14

If the diagonals of the parallelogram **ABCD** intersect at **O**,
Then **AO = OC** and **BO = OD**.

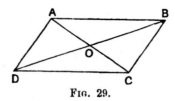

Fig. 29.

Theorem 15

If **AB** is equal and parallel to **CD**,
Then **AC** is equal and parallel to **BD**.

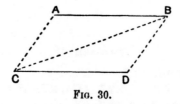

Fig. 30.

DEFINITIONS.—A *parallelogram* is a four-sided figure whose opposite sides are parallel.

A *rectangle* is a parallelogram, *one* angle of which is a right angle.

A *square* is a rectangle, having two adjacent sides equal.

A *rhombus* is a parallelogram, having two adjacent sides equal, but none of its angles right angles.

A *trapezium* is a four-sided figure with one pair of opposite sides parallel.

CONGRUENT TRIANGLES (Second Section), PARALLELOGRAMS, SQUARES, Etc.

EXERCISE V

1. Prove that all the sides of a rhombus are equal.
2. Prove that the diagonals of a rectangle are equal.
3. Prove that the diagonals of a rhombus intersect at right angles.
4. Prove that the diagonals of a square are equal and cut at right angles.
5. The diagonals of the rectangle **ABCD** meet at **O**; $\angle$ **BOC** = 44°; calculate $\angle$ **OAD**.
6. Prove that a quadrilateral, whose opposite sides are equal, is a parallelogram.
7. **ABCD** is a rhombus; $\angle$ **ABC** = 56°; calculate $\angle$ **ACD**.
8. **ABCD** is a parallelogram; prove that **B** and **D** are equidistant from **AC**.
9. **X** is the mid-point of a chord **AB** of a circle, centre **O**; prove $\angle$ **OXA** = 90°.
10. The diagonals of the parallelogram **ABCD** cut at **O**; any line through **O** cuts **AB**, **CD** at **X**, **Y**; prove **XO** = **OY**.
11. Two straight lines **POQ**, **ROS** cut at **O**; if **PQ** = **RS** and **PR** = **QS**, prove $\angle$ **RPO** = $\angle$ **QSO**.
12. In the quadrilateral **ABCD**, **AB** = **CD** and **AC** = **BD**; prove that **AD** is parallel to **BC**.
13. **E** is a point inside the square **ABCD**; a square **AEFG** is described on the same side of **AE** as **D**; prove **BE** = **DG**.
14. **ABC** is any triangle; **BY**, **CZ** are lines parallel to **AC**, **AB** cutting a line through **A** parallel to **BC** in **Y**, **Z**; prove **AY** = **AZ**.
15. **ABCD** is a parallelogram; **P** is the mid-point of **BC**; **DP** and **AB** are produced to meet at **Q**; prove **AQ** = 2**AB**.
16. **ABCD**, **ABXY** are two parallelograms; **BC** and **BX** are different lines; prove that **DCXY** is a parallelogram.
17. Two unequal circles, centres **A**, **B**, intersect at **X**, **Y**; prove that **AB** bisects **XY** at right angles.
18. The diagonals of a square **ABCD** cut at **O**; from **AB** a part **AK** is cut off equal to **AO**; prove $\angle$ **AOK** = 3 $\angle$ **BOK**.

19. **ABCD** is a straight line such that **AB = BC = CD**; **BCPQ** is a rhombus; prove that **AQ** is perpendicular to **DP**.
20. **ABCD** is a parallelogram; the bisector of ∠ **ABC** cuts **AD** at **X**; the bisector of ∠ **BAD** cuts **BC** at **Y**; prove **XY = CD**.
21. **ABCD** is a parallelogram such that the bisectors of ∠s **DAB**, **ABC** meet on **CD**; prove **AB = 2BC**.
22. In △**ABC**, ∠**BAC** = 90°; **BADH**, **ACKE** are squares outside the triangle; prove that **HAK** is a straight line.
23. The diagonals of the rectangle **ABCD** cut at **O**; **AO > AB**; the circle, centre **A**, radius **AO** cuts **AB** produced at **E**; if ∠**AOB** = 4∠**BOE**, calculate ∠**BAC**.

----

24. **ABC** is an equilateral triangle; a line parallel to **AC** cuts **BA**, **BC** at **P**, **Q**; **AC** is produced to **R** so that **BQ = CR**; prove that **PR** bisects **CQ**.
25. **P** is one point of intersection of two circles, centres **A**, **B**; **AQ**, **BR** are radii parallel to and in the same sense as **BP**, **AP**; prove that **QPR** is a straight line.
26. In △**ABC**, ∠**BAC** = 90°; **ABPQ**, **ACRS**, **BCXY** are squares outside **ABC**; prove that (i) **BQ** is parallel to **CS**; (ii) **BR** is perpendicular to **AX**.
27. **ABC** is a triangle; the bisectors of ∠s **ABC**, **ACB** meet at **I**; prove **IA** bisects ∠**BAC**. [From **I** drop perpendiculars to **AB**, **BC**, **CA**.]
28. In △**ABC**, ∠**BAC** = 90°; **BCPQ**, **ACHK** are squares outside **ABC**; **AC** cuts **PH** at **D**; prove **AB = 2CD** and **PD = DH**.
29. In △**ABC**, **AB = AC**; **P** is any point on **BC**; **PX**, **PY** are the perpendiculars from **P** to **AB**, **AC**; **CD** is the perpendicular from **C** to **AB**; prove **PX + PY = CD**.
30. In △**ABC**, **AB = AC**; **P** is a variable point on **BC**; **PQ**, **PR** are lines parallel to **AB**, **AC** cutting **AC**, **AB** at **Q**, **R**; prove that **PQ + PR** is constant.
31. **H**, **K** are the mid-points of the sides **AB**, **AC** of △**ABC**; **HK** is joined and produced to **X** so that **HK = KX**; prove that (i) **CX** is equal and parallel to **BH**; (ii) **HK** = ½**BC** and **HK** is parallel to **BC**.

RIDERS ON BOOK II

AREAS

Theorem 16

(i) If **ABCD** and **ABPQ** are parallelograms on the same base and between the same parallels, their areas are equal.

(ii) If **BH** is the height of the parallelogram **ABCD**,
$$\text{area of } \mathbf{ABCD} = \mathbf{AB} \cdot \mathbf{BH}.$$

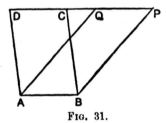

Fig. 31.

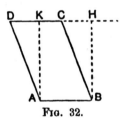

Fig. 32.

Theorem 17

If **AD** is an altitude of the triangle **ABC**,
$$\text{area of } \mathbf{ABC} = \tfrac{1}{2}\, \mathbf{AD} \cdot \mathbf{BC}.$$

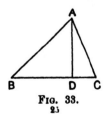

Fig. 33.

Theorem 18

(i) If **ABC** and **ABD** are triangles on the same base and between the same parallels, their areas are equal.

(ii) If the triangle **ABC**, **ABD** are of equal area and lie on the same side of the common base **AB**, they are between the same parallels, *i.e.* **CD** is parallel to **AB**.

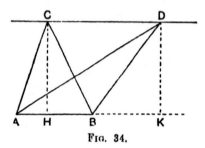

Fig. 34.

Theorem 19(1)

If the triangle **ABC** and the parallelogram **ABXY** are on the same base **AB** and between the same parallels,

area of **ABC** = ½ area of **ABXY**.

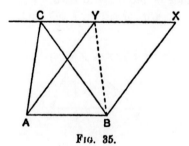

Fig. 35.

AREAS

THEOREM 19(2)

(i) Triangles (or parallelograms) on equal bases and between the same parallels are equal in area.

(ii) Triangles (or parallelograms) of equal area, which are on equal bases in the same straight line and on the same side of it, are between the same parallels.

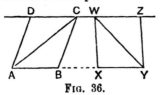

FIG. 36.

MENSURATION THEOREMS

(i) If the lengths of the parallel sides of a trapezium are a inches and b inches, and if their distance apart is h inches,

area of trapezium $= \tfrac{1}{2} h (a+b)$ sq. inches.

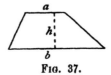

FIG. 37.

(ii) If the lengths of the sides of a triangle are a, b, c inches and if $s = \tfrac{1}{2}(a+b+c)$,

area of triangle $= \sqrt{s(s-a)(s-b)(s-c)}$ sq. inches.

AREAS

TRIANGLES, PARALLELOGRAMS, ETC.

EXERCISE VI

In Fig. 38, **AD, BE, CF** are altitudes of the triangle **ABC**.

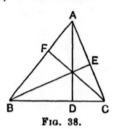

Fig. 38.

1. In △**ABC**, ∠**ABC** = 90°, **AB** = 3″, **BC** = 5″; find area of **ABC**.
2. In Fig. 38, **AD** = 7″, **BC** = 5″; find area of **ABC**.
3. In Fig. 38, **BE** = 5″, **CF** = 6″, **AB** = 4″; find **AC**.
4. In Fig. 38, **AD** = $6x″$, **BE** = $4x″$, **CF** = $3x″$, and the perimeter of **ABC** is 18″. Find **BC**.
5. In quad. **ABCD**, **AB** = 12″, **BC** = 1″, **CD** = 9″, **DA** = 8″, ∠**ABC** = ∠**ADC** = 90°; find the area of **ABCD**.
6. In quad. **ABCD**, **AC** = 8″, **BD** = 11″, and **AC** is perpendicular to **BD**; find the area of **ABCD**.
7. Find the area of a triangle whose sides are 3″, 4″, 5″.

In Fig. 39, **ABCD** is a parallelogram; **AP, AQ** are the perpendiculars to **BC, CD**.

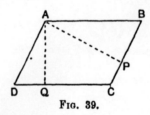

Fig. 39.

8. In Fig. 39, **AB** = 7″, **AQ** = 3″; find the area of **ABCD**.
9. In Fig. 39, **AB** = 5″, **AD** = 4″, **AP** = 6″; find **AQ**.
10. In Fig. 39, **AP** = 3″, **AQ** = 2″, and perimeter of **ABCD** is 20″; find its area.

AREAS

11. In quad. ABCD, BC = 8″, AD = 3″, and BC is parallel to AD; if the area of △ABC is 18 sq. in., find the area of △ABD.
12. In quad. ABCD, AB = 5″, BC = 3″, CD = 2″, ∠ABC = ∠BCD = 90°; find the area of ABCD.
13. In Fig. 38, AB = 8″, AC = 6″, BE = 5″; find CF.
14. The area of △ABC is 36 sq. cms., AB = 8 cms., AC = 9 cms., D is the mid-point of BC; find the lengths of the perpendiculars from D to AB, AC.
15. In the parallelogram ABCD, AB = 8″, BC = 5″; the perpendicular from A to CD is 3″; find the perpendicular from B to AD.
16. Find the area of a rhombus whose diagonals are 5″, 6″.
17. In △ABC, ∠ABC = 90°, AB = 6″, BC = 8″, CA = 10″; D is the mid-point of AC. Calculate the lengths of the perpendiculars from B to AC and from A to BD.
18. On an Ordnance Map, scale 6 inches to the mile, a football field is approximately a square measuring $\frac{1}{2}$ inch each way. Find the area of the field in acres, correct to $\frac{1}{10}$ acre.
19. Fig. 40 represents on a scale of 1″ to the foot a trough and the depth of water in it. The water is running at 4 miles an hour; find the number of gallons which pass any point in a minute, to nearest gallon, taking 1 cub. ft. = $6\frac{1}{4}$ gallons.

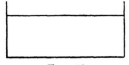

Fig. 40.

20. Fig. 41 represents on a scale of 1 cm. to 100 yds. the plan of a field; find its area in acres correct to nearest acre.

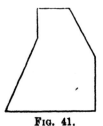

Fig. 41.

21. Fig. 42 represents the plan and elevation of a box on a scale of 1 cm. to 1 ft.
    (i) Find the volume of the box.
    (ii) Find the *total* area of its surface.

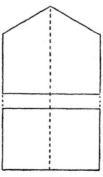

Fig. 42.

22. The diagram (Fig. 43), not drawn to scale, represents the plan of an estate of 6⅔ acres. The measurements given are in inches. On what scale (inches to the mile) is it drawn? The dotted line PQ divides the estate in half; find AQ.

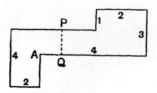

Fig. 43.

23. Find the area of ABCD (Fig. 44) in terms of x, y, p, q, r.

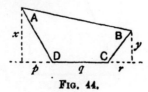

Fig. 44.

AREAS

24. ABC is inscribed in a rectangle (Fig. 45); find the area of ABC in terms of p, q, r, s.

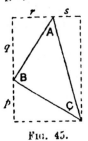

Fig. 45.

25. In Fig. 46 $\angle ABC = \angle BCD = 90°$. Find the length of the perpendicular from C to AD in terms of p, q, r.

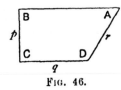

Fig. 46.

26. In Fig. 47 OB is a square, side $4''$; $OA = 12''$, $OC = 6''$. Calculate areas of $\triangle OAB$, $\triangle OBC$, $\triangle AOC$, and prove that ABC is a straight line.

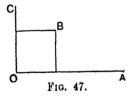

Fig. 47.

27. In $\triangle AOB$, $OA = a$, $OB = b$, $\angle AOB = 90°$; P is a point on AB; PH, PK are the perpendiculars from P to OA, OB; $PH = x$, $PK = y$; prove $\dfrac{x}{a} + \dfrac{y}{b} = 1$.

28. P, Q are points on the sides AB, AD of the rectangle ABCD; $AB = x$, $AD = y$, $PB = e$, $QD = f$. Calculate area of PCQ in terms of e, f, x, y.

29. The area of a rhombus is 25 sq. cms., and one diagonal is half the other; calculate the length of each diagonal.

32   CONCISE GEOMETRY

30. Find the area of the triangles whose vertices are:
      (i) (2, 1); (2, 5); (4, 7).
      (ii) (3, 2); (5, 4); (4, 8).
      (iii) (1, 1); (5, 2); (6, 5).
      (iv) (0, 0); (a, o); (b, c).
      (v) (0, 0); (a, b); (c, d).
31. Find the area of the quadrilaterals whose vertices are:
      (i) (0, 0); (3, 2); (1, 5); (0, 7).
      (ii) (1, 3); (3, 2); (5, 5); (2, 7).
32. Find in acres the areas of the fields of which the following field-book measurements have been taken:

		YARDS				YARDS	
		to D				to D	
		250				300	
(1)	to C 80	200				220	50 to E
		150	40 to E	(2)	to C 60	200	
	to B 50	100			to B 100	100	
						50	80 to F
		From A				From A	

33. Find from the formula [page 27] the area of the triangles whose sides are (i) 5 cms., 6 cms., 7 cms.
                (ii) 8″, 15″, 19″.

Find also in each case the greatest altitude.

34. The sides of a triangle are 7″, 8″, 10″. Calculate its shortest altitude.

35. AX, BY are altitudes of the triangle ABC; if AC = 2BC, prove AX = 2BY.
36. ABC is a △; a line parallel to BC cuts AB, AC at P, Q; prove △APC = △AQB.
37. Two lines AOB, COD intersect at O; if AC is parallel to BD, prove △AOD = △BOC.
38. The diagonals AC, BD of ABCD are at right angles, prove that area of ABCD = ½ AC . BD.
39. The diagonals of the quad. ABCD cut at O; if △AOB = △AOD, prove △DOC = △BOC.
40. In the triangles ABC, XYZ, AB = XY, BC = YZ, ∠ABC + ∠XYZ = 180°, prove △ABC = △XYZ.

AREAS

41. P is any point on the median AD of △ABC; prove △APB = △APC.
42. ABCD is a quadrilateral; lines are drawn through A, C parallel to BD, and through B, D parallel to AC; prove that the area of the parallelogram so obtained equals twice the area of ABCD.
43. ABCD is a parallelogram; P is any point on AD; prove that △PAB + △PCD = △PBC.
44. ABC is a straight line; O is a point outside it; prove
$$\frac{\triangle OAB}{\triangle OBC} = \frac{AB}{BC}.$$
45. ABCD is a parallelogram; P is any point on BC; DQ is the perpendicular from D to AP; prove that the area of ABCD = DQ . AP.
46. ABCD is a parallelogram; P is any point on BD; prove △PAB = △PBC.
47. ABCD is a parallelogram; a line parallel to BD cuts BC, DC at P, Q; prove △ABP = △ADQ.
48. AOB is an angle; X is the mid-point of OB; Y is the mid-point of AX; prove △AOY = △BXY.
49. If in Fig. 48, AC is perpendicular to BD, prove area of ABCD = ½AC . BD.

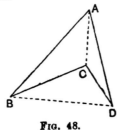

Fig. 48.

50. ABCD is a quadrilateral; a line through D parallel to AC meets BC produced at P; prove that △ABP = quad. ABCD.
51. ABCD is a quadrilateral; E, F are the mid-points of AB, CD; prove that △ADE + △CBF = △BCE + △ADF.
52. The diagonals of a quadrilateral divide it into four triangles of equal area; prove that it is a parallelogram.
53. ABCD and PQ are parallel lines; AB = BC = CD = PQ; PC cuts BQ at O; prove quad. ADQP = 8 △OBC.

54. **X, Y** are the mid-points of the sides **AB, AC** of △**ABC**; prove that △**XBY** = △**XCY** and deduce that **XY** is parallel to **BC**.
55. Two parallelograms **ABCD, AXYZ** of equal area have a common angle at **A**; **X** lies on **AB**; prove **DX, YC** are parallel.
56. The sides **AB, BC** of the parallelogram **ABCD** are produced to any points **P, Q**; prove △**PCD** = △**QAD**.
57. **ABC** is a △; **D, E** are the mid-points of **BC, CA**; **Q** is any point in **AE**; the line through **A** parallel to **QD** cuts **BD** at **P**; prove **PQ** bisects △**ABC**.
58. The medians **BE, CF** of △**ABC** intersect at **G**; prove that △**BGC** = △**BGA** = △**AGC**.
59. In Fig. 49, the sides of △**ABC** are equal and parallel to the sides of △**XYZ**; prove that **BAXY** + **ACZX** = **BCZY**.

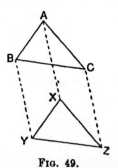

Fig. 49.

60. **ABP, AQB** are equivalent triangles on opposite sides of **AB**; prove **AB** bisects **PQ**.
61. **ABCD** is a parallelogram; any line through **A** cuts **DC** at **Y** and **BC** produced at **Z**; prove △**BCY** = △**DYZ**.
62. In Fig. 50, **PR** is equal and parallel to **AB**; **PQAT** and **CQRS** are parallelograms; prove they are equivalent.

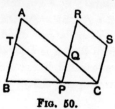

Fig. 50.

AREAS

63. BE, CF are medians of the triangle ABC and cut at G; prove △BGC = quad. AEGF.
64. ABC, ABD are triangles on the same base and between the same parallels; BC cuts AD at O; a line through O parallel to AB cuts AC, BD at X, Y; prove XO = OY.
65. In Fig. 51, APQR is a square; prove $\dfrac{1}{AP} = \dfrac{1}{AB} + \dfrac{1}{AC}$.

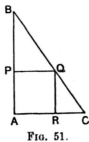

Fig. 51.

66. ABCD is a quadrilateral; AB is parallel to CD; P is the mid-point of BC; prove ABCD = 2△APD.
67. ABCD is a parallelogram; DC is produced to P; AP cuts BD at Q; prove △DQP − △AQB = △BCP.
68. In Fig. 52, ABCD is divided into four parallelograms; prove POSD = ROQB.

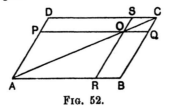

Fig. 52.

69. In Fig. 52, prove △APR + △ASQ = △ABD.
70. ABC is a △; any three parallel lines AX, BY, CZ meet BC, CA, AB produced where necessary at X, Y, Z; prove △AYZ = △BZX = △CXY.
71. In ex. 70, prove △XYZ = 2△ABC.
72. ABCD is a parallelogram; AB is produced to E; P is any point within the angle CBE; prove △PAB + △PBC = △PBD.

73*. ABC is a △; ACPQ, BCRS are parallelograms outside ABC; QP, SR are produced to meet at O; ABXY is a parallelogram such that BX is equal and parallel to OC; prove that ACPQ + BCRS = ABXY.

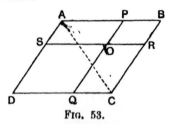

Fig. 53.

74*. In Fig. 53, ABCD is divided into four parallelograms, prove that SOQD − BPOR = 2△AOC.
75*. P is a variable point inside a fixed equilateral triangle ABC; PX, PY, PZ are the perpendiculars from P to BC, CA, AB; prove that PX + PY + PZ is constant.
76*. In △ABC, ∠ABC = 90°; DBC is an equilateral triangle outside ABC; prove △ADC − △DBC = ½△ABC.
77*. In △ABC, ∠BAC = 90°; X, Y, Z are points on AB, BC, CA such that AXYZ is a rectangle and AX = ¼AB; prove AXYZ = ⅜△ABC.
78*. Two fixed lines BA, DC meet when produced at O; E, F are points on OB, OD such that OE = AB, OF = CD; P is a variable point in the angle BOD such that △PAB + △PCD is constant; prove that the locus of P is a line parallel to EF.
79*. G, H are the mid-points of the diagonals AC, BD of the quadrilateral ABCD; AB and DC are produced to meet at P; prove quad. ABCD = 4△PGH.

PYTHAGORAS' THEOREM

Theorem 20

If, in the triangle **ABC**, $\angle$ **BAC** $= 90°$,
Then **BA**^{2} + **AC**^{2} = **BC**^{2}.

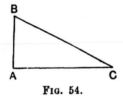

Fig. 54.

Theorem 21

If, in the triangle **ABC**, **BA**^{2} + **AC**^{2} = **BC**^{2},
Then $\angle$ **BAC** $= 90°$.

PYTHAGORAS' THEOREM

EXERCISE VII

1. In Fig. 54, AB = 5″, AC = 12″, calculate BC.
2. In Fig. 54, AC = 6″, BC = 10″, calculate AB.
3. In Fig. 54, AB = 7″, BC = 9″, calculate AC.
4. In △ABC, AB = AC = 9″, BC = 8″, calculate area of △ABC.
5. In △ABC, AB = AC = 13″, BC = 10″, calculate the length of the altitude BE.
6. Find the side of a rhombus whose diagonals are 6, 10 cms.
7. A kite at P, flown by a boy at Q, is vertically above a point R on the same level as Q; if PQ = 505′, QR = 456′, find the height of the kite.
8. In △ABC, AC = 3″, AB = 8″, ∠ACB = 90°; find the length of the median AD.
9. AD is an altitude of △ABC; AD = 2″, BD = 1″, DC = 4″; prove ∠BAC = 90°.
10. ABCD is a parallelogram; AC = 13″, BD = 5″, ∠ABD = 90°; calculate area of ABCD.
11. A gun, whose effective range is 9000 yards, is 5000 yards from a straight railway; what length of the railway is commanded by the gun?
12. The lower end of a 20-foot ladder is 10 feet from a wall; how high up the wall does the ladder reach? How much closer must it be put to reach one foot higher?
13. An aeroplane heads due North at 120 miles an hour in an east wind blowing at 40 miles an hour; find the distance travelled in ten minutes.
14. A ship is steaming at 15 knots and heading N.W.; there is a 6-knot current setting N.E.; how far will she travel in one hour?
15. AB, AC are two roads meeting at right angles; AB = 110 yards, AC = 200 yards; P starts from B and walks towards A at 3 miles an hour; at the same moment Q starts from C and walks towards A at 4 miles an hour. How far has P walked before he is within 130 yards of Q?
16. Find the distance between the points (1, 2), (5, 4).

PYTHAGORAS' THEOREM

17. Prove that the points (5, 11), (6, 10), (7, 7) lie on a circle whose centre is (2, 7); and find its radius.
18. The parallel sides of an isosceles trapezium are 5", 11", and its area is 32 sq. inches; find the lengths of the other sides.
19. In $\triangle ABC$, $\angle ABC = 90°$, $\angle ACB = 60°$, $AC = 8"$; find AB.
20. In $\triangle ABC$, $\angle ABC = 90°$, $\angle ACB = 60°$, $AB = 5"$; find BC.
21. In Fig. 55, $AB = 2"$, $BC = 4"$, $CD = 1"$; find AD.

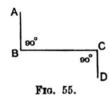

FIG. 55.

22. In quadrilateral ABCD, $AB = 5"$, $BC = 12"$, $CD = 7"$; $\angle ABC = \angle BCD = 90°$; P, Q are points on BC such that $\angle APD = 90° = \angle AQD$; calculate BP, BQ.
23. In Fig. 56, $AC = CB = 12"$, $CD = 8"$, $\angle ACD = 90°$; find radius of circular arc.

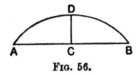

FIG. 56.

24. Prove that the triangle whose sides are $x^2 + y^2$, $x^2 - y^2$, $2xy$ is right-angled.
25. AD is an altitude of the triangle ABC; $BD = x^2$, $DC = y^2$, $AD = xy$; prove that $\angle BAC = 90°$.
26. AD is an altitude of $\triangle ABC$, $\angle BAC = 90°$; $AD = 4"$, $CD = 3"$; calculate AB.
27. AD, BC are two vertical poles, D and C being the ends on the ground, which is level; $AC = 12'$, $AB = 10'$, $BC = 3'$; calculate AD.
28. AD, BC are the parallel sides of the trapezium ABCD; $AB = 6$, $BC = 9$, $CD = 5$, $AD = 14$; find the area of ABCD.
29. In $\triangle ABC$, $AB = AC = 10"$, $BC = 8"$; find the radius of the circle which passes through A, B, C.

30. In $\triangle ABC$, $AB = 4''$, $BC = 5''$, $\angle ABC = 45°$; calculate AC.
31. In $\triangle ABC$, $AB = 8''$, $BC = 3''$, $\angle ABC = 60°$; calculate AC.
32. A regular polygon of n sides is inscribed in a circle, radius r; its perimeter is p; prove that its area is $\dfrac{p}{2}\sqrt{\left(r^2 - \dfrac{p^2}{4n^2}\right)}$.

    Hence, assuming that the circumference of a circle of radius r is $2\pi r$, prove that the area of the circle is πr^2.
33. The slant side of a right circular cone is $10''$, and the diameter of its base is $8''$; find its height.
34. Find the diagonal of a cube whose edge is $5''$.
35. A room is 20 feet long, 16 feet wide, 8 feet high; find the length of a diagonal.
36. A piece of wire is bent into three parts AB, BC, CD each of the outer parts being at right angles to the plane containing the other two; $AB = 12''$, $BC = 6''$, $CD = 12''$; find the distance of A from D.
37. A hollow sphere, radius $8''$, is filled with water until the surface of the water is within $3''$ of the top. Find the radius of the circle formed by the water-surface.
38. A circular cone is of height h feet, and the radius of its base is r feet; prove that the radius of its circumscribing sphere is $\dfrac{h}{2} + \dfrac{r^2}{2h}$ feet.
39. A pyramid of height $8''$ stands on a square base each edge of which is $1'$. Find the area of the sides and the length of an edge.
40*. ABCD is a rectangle; $AB = 6''$, $BC = 8''$; it is folded about BD so that the planes of the two parts are at right angles. Find the new distance of A from C.

---

41. AD is an altitude of the equilateral triangle ABC; prove that $AD^2 = \tfrac{3}{4}BC^2$.
42. In $\triangle ABC$, $\angle ACB = 90°$; CD is an altitude; prove $AC^2 + BD^2 = BC^2 + AD^2$.
43. ABN, PQN are two perpendicular lines; prove that $PA^2 + QB^2 = PB^2 + QA^2$.
44. The diagonals AC, BD of the quadrilateral ABCD are at right angles; prove that $AB^2 + CD^2 = AD^2 + BC^2$.

45. If in the quadrilateral **ABCD**, $\angle$ **ABC** = $\angle$ **ADC** = $90°$; prove that $AB^2 - AD^2 = CD^2 - CB^2$.
46. **P** is a point inside a rectangle **ABCD**; prove that $PA^2 + PC^2 = PB^2 + PD^2$. Is this true if **P** is outside **ABCD**?
47. In $\triangle$**ABC**, $\angle$ **BAC** = $90°$; **H**, **K** are the mid-points of **AB**, **AC**; prove that $BK^2 + CH^2 = \tfrac{5}{4}BC^2$.
48. **ABCD** is a rhombus; prove that $AC^2 + BD^2 = 2AB^2 + 2BC^2$.
49. In the quadrilateral **ABCD**, $\angle$ **ACB** = $\angle$ **ADB** = $90°$; **AH**, **BK** are drawn perpendicular to **CD**; prove $DH^2 + DK^2 = CH^2 + CK^2$.
50. **PX, PY, PZ, PW** are the perpendiculars from a point **P** to the sides of the rectangle **ABCD**; prove that $PA^2 + PB^2 + PC^2 + PD^2 = 2(PX^2 + PY^2 + PZ^2 + PW^2)$.
51. In $\triangle$**ABC**, $\angle$ **BAC** = $90°$ and **AC** = 2**AB**; **AC** is produced to **D** so that **CD** = **AB**; **BCPQ** is the square on **BC**; prove **BP** = **BD**.
52. **AD** is an altitude of $\triangle$**ABC**; **P, Q** are points on **AD** produced such that **PD** = **AB** and **QD** = **AC**; prove **BQ** = **CP**.
53. In $\triangle$**ABC**, $\angle$ **BAC** = $90°$; **AD** is an altitude; prove
$$AD = \frac{AB \cdot AC}{BC}.$$
54. In $\triangle$**ABC**, $\angle$ **BAC** = $90°$; **AX** is an altitude; use Fig. 24, page 15, and the proof of Pythagoras' theorem to show that $BA^2 = BX \cdot BC$; and deduce that $\dfrac{AB^2}{AC^2} = \dfrac{BX}{CX}$.
55. In $\triangle$**ABC**, $\angle$ **BAC** = $90°$; **AD** is an altitude; prove that $AD^2 = BD \cdot DC$.
56. **ABC** is an equilateral triangle; **D** is a point on **BC** such that $BC = 3BD$; prove $AD^2 = \tfrac{7}{9}AB^2$.
57. **ABC** is an equilateral triangle; **D, E** are the mid-points of **BC, CD**; prove $AE^2 = 13EC^2$.
58. In the $\triangle$**ABC**, $AB = AC = 2BC$; **BE** is an altitude; prove that $AE = 7EC$.
59. **O** is any point inside $\triangle$**ABC**; **OP, OQ, OR** are the perpendiculars to **BC, CA, AB**; prove $BP^2 + CQ^2 + AR^2 = PC^2 + QA^2 + RB^2$.
60. **AD** is an altitude of $\triangle$**ABC**; **E** is the mid-point of **BC**; prove $AB^2 \sim AC^2 = 2BC \cdot DE$.

61. Fig. 57 shows a square of side $a+b$ divided up; use area formulæ to prove Pythagoras' theorem $a^2 + b^2 = c^2$.

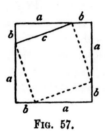

Fig. 57.

62*. ABC is a straight line; ABXY, BCPQ are squares on the same side of AC; prove $PX^2 + CY^2 = 3(AB^2 + BC^2)$.

63*. The diagonal AC of the rhombus ABCD is produced to any point P; prove that $PA \cdot PC = PB^2 - AB^2$.

64*. The diagonal AC of the square ABCD is produced to P so that $PC = BC$; prove $PB^2 = PA \cdot AC$.

65*. In $\triangle ABC$, $\angle BAC = 90°$; BCXY, ACPQ, ABRS are squares outside ABC; prove $PX^2 + RY^2 = 5BC^2$.

EXTENSIONS OF PYTHAGORAS' THEOREM

THEOREM 22

In $\triangle ABC$, if $\angle BAC$ is obtuse and if CN is the perpendicular to BA produced,

then $BC^2 = BA^2 + AC^2 + 2BA \cdot AN$.

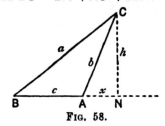

FIG. 58.

THEOREM 23

In $\triangle ABC$, if $\angle BAC$ is acute, and if CN is the perpendicular to AB or AB produced,

then $BC^2 = BA^2 + AC^2 - 2BA \cdot AN$.

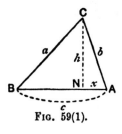
FIG. 59(1).

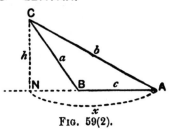
FIG. 59(2).

THEOREM 24

In $\triangle ABC$, if AD is a median,

then $AB^2 + AC^2 = 2AD^2 + 2DB^2$.

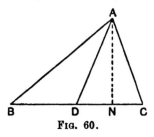

FIG. 60.

EXTENSIONS OF PYTHAGORAS' THEOREM

EXERCISE VIII

1. Find by calculation which of the following triangles are obtuse-angled, their sides being as follows :—(i) 4, 5, 7 ; (ii) 7, 8, 11 ; (iii) 8, 9, 12 ; (iv) 15, 16, 22.
2. Each of the sides of an acute-angled triangle is an exact number of inches; two of them are 12″, 15″; what is the greatest length of the third side ?
3. In △ABC, BC = 6, CA = 3, AB = 4 ; CN is an altitude ; calculate AN and CN.
4. In △ABC, BC = 8, CA = 9, AB = 10 ; CN is an altitude ; calculate AN and CN.
5. In △ABC, BC = 7, CA = 13, AB = 10 ; CN is an altitude ; calculate AN, BN, CN.
6. Find the area of the triangle whose sides are 9″, 10″, 11″.
7. ABCD is a parallelogram ; AB = 5″, AD = 3″; the projection of AC on AB is 6″; calculate AC.
8. In △ABC, AC = 8 cms., BC = 6 cms., ∠ACB = 120°; calculate AB.
9. In △ABC, AB = 8 cms., AC = 7, BC = 3 ; prove ∠ABC = 60°.
10. The sides of a triangle are 23, 27, 36 ; is it obtuse-angled?
11. In △ABC, AB = 9″, AC = 11″, ∠BAC > 90°; prove BC > 14″.
12. In △ABC, AB = 14″, BC = 10″, CA = 6″; prove ∠ACB = 120°.
13. The sides of a △ are 4, 7, 9 ; calculate the length of the shortest median.
14. Find the lengths of the medians of a triangle whose sides are 6, 8, 9 cms.
15. The sides of a parallelogram are 5 cms., 7 cms., and one diagonal is 8 cms. ; find the length of the other.
16. AD is a median of the △ABC, AB = 6, AC = 8, AD = 5 ; calculate BC.
17. In △ABC, AB = 4, BC = 5, CA = 8 ; BC is produced to D so that CD = 5 ; calculate AD.

18. ABC is an equilateral triangle ; BC is produced to D so that BC = CD ; prove $AD^2 = 3AB^2$.

19. In $\triangle ABC$, $AB = AC$; CD is an altitude; prove that $BC^2 = 2AB \cdot BD$.

20. AB and DC are the parallel sides of the trapezium ABCD; prove that $AC^2 + BD^2 = AD^2 + BC^2 + 2AB \cdot DC$.

21. BE, CF are altitudes of the triangle ABC; prove that $AF \cdot AB = AE \cdot AC$.

22. ABCD is a parallelogram; prove that $AC^2 + BD^2 = 2AB^2 + 2BC^2$.

23. ABCD is a rectangle; P is any point in the same or any other plane; prove that $PA^2 + PC^2 = PB^2 + PD^2$.

24. In $\triangle ABC$, $AB = AC$; AB is produced to D so that $AB = BD$; prove $CD^2 = AB^2 + 2BC^2$.

25. In $\triangle ABC$, D, E are the mid-points of CB, CA; prove that $4(AD^2 - BE^2) = 3(CA^2 - CB^2)$.

26. In $\triangle ABC$, $\angle ACB = 90°$; AB is trisected at P, Q; prove that $PC^2 + CQ^2 + QP^2 = \frac{2}{3}AB^2$.

27. The base BC of $\triangle ABC$ is trisected at X, Y; prove that $AX^2 + AY^2 + 4XY^2 = AB^2 + AC^2$.

28. The base BC of $\triangle ABC$ is trisected at X, Y; prove that $AB^2 - AC^2 = 3(AX^2 - AY^2)$.

29. AD, BE, CF are the medians of $\triangle ABC$; prove that $4(AD^2 + BE^2 + CF^2) = 3(AB^2 + BC^2 + CA^2)$.

30. ABCD is a quadrilateral; X, Y are the mid-points of AC, BD; prove that $AB^2 + BC^2 + CD^2 + DA^2 = AC^2 + BD^2 + 4XY^2$.

31*. ABC is a triangle; ABPQ, ACXY are squares outside ABC; prove that $BC^2 + QY^2 = AP^2 + AX^2$.

32*. ABC is a triangle; D is a point on BC such that $p \cdot BD = q \cdot DC$; prove that $p \cdot AB^2 + q \cdot AC^2 = (p+q) AD^2 + p \cdot BD^2 + q \cdot DC^2$.

33*. AB is a diameter of a circle; PQ is any chord parallel to BA; O is any point on AB; prove that $OP^2 + OQ^2 = OA^2 + OB^2$.

34*. ABCD is a tetrahedron; $\angle BAC = \angle CAD = \angle DAB = 90°$; prove that BCD is an acute-angled triangle.

RELATIONS BETWEEN SEGMENTS OF A STRAIGHT LINE

EXERCISE IX

1. A straight line **AB** is bisected at **O**; **P** is any point on **AO**; prove $PO = \frac{1}{2}(PB - PA)$.

2. A straight line **AB** is bisected at **O** and produced to **P**; prove that $OP = \frac{1}{2}(AP + BP)$.

3. A straight line **AB** is bisected at **O** and produced to **P**; prove that $PA^2 + PB^2 = 2PO^2 + 2AO^2$.

4. **ABCD** is a straight line; **X, Y** are the mid-points of **AB, CD**; prove that $AD + BC = 2XY$.

5. **AB** is bisected at **O** and produced to **P**; prove that $AO \cdot AP = OB \cdot BP + 2AO^2$.

6. **AD** is trisected at **B, C**; prove that $AD^2 = AB^2 + 2BD^2$.

7. **APB** is a straight line; prove that $AB^2 + AP^2 = 2AB \cdot AP + PB^2$.

8. **AB** is bisected at **C** and produced at **P**; prove that $AP^2 = 4AC \cdot CP + BP^2$.

9. **ABCD** is a straight line; if $AB = CD$, prove that $AD^2 + BC^2 = 2AB^2 + 2BD^2$.

10. **X** is a point on **AB** such that $AB \cdot BX = AX^2$; prove that $AB^2 + BX^2 = 3AX^2$.

11. **C** is a point on **AB** such that $AB \cdot BC = AC^2$; prove that $AC \cdot BC = AC^2 - BC^2$.

12. **X** is a point on **AB** such that $AB \cdot BX = AX^2$; **O** is the mid-point of **AX**; prove that $OB^2 = 5 \cdot OA^2$.

13. **AB** is bisected at **O** and produced to **P** so that $OB \cdot OP = BP^2$; prove that $PA^2 = 5PB^2$.

14. **AB** is bisected at **C** and produced to **D** so that $AD^2 = 3CD^2$; **BC** is bisected at **P**; prove that $PD^2 = 3PB^2$.

15. **AB** is produced to **P** so that $PA^2 = 4PB^2 + AB^2$; prove that $\dfrac{PA}{PB} = \dfrac{5}{2}$.

16. **ACBD** is a straight line such that $\dfrac{AC}{CB}=\dfrac{AD}{BD}$; **O** is the mid-point of **AB**; prove that

 (i) DA.DB = DC.DO.
 (ii) AB.CD = 2AD.CB.
 (iii) OB^2 = OC.OD.
 (iv) $\dfrac{1}{AC}+\dfrac{1}{AD}=\dfrac{2}{AB}$.

INEQUALITIES

THEOREM 26

In the triangle **ABC**,
    (i) If **AC** > **AB**, then ∠ **ABC** > ∠ **ACB**.
    (ii) If ∠ **ABC** > ∠ **ACB**, then **AC** > **AB**.

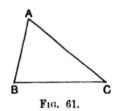

FIG. 61.

THEOREM 27

If **ON** is the perpendicular from any point **O** to a line **AB**, and if **P** is any point on **AB**,
    then **ON** < **OP**.

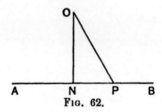

FIG. 62.

THEOREM 28

If **ABC** is a triangle, **BA** + **AC** > **BC**.

INEQUALITIES

EXERCISE X

1. The bisectors of the angles ABC, ACB of △ABC meet at I; if AB > AC, prove that IB > IC.
2. AD is a median of △ABC; if BC < 2AD, prove that ∠BAC < 90°.
3. ABC is an equilateral triangle; P is any point on BC; prove AP > BP.
4. In △ABC, the bisector of ∠BAC cuts BC at D; prove BA > BD.
5. AD is a median of △ABC; if AB > AC, prove that ∠BAD < ∠CAD.
6. In △ABC, AB = AC; BC is produced to any point D; P is any point on AB; DP cuts AC at Q; prove AQ > AP.
7. In the quadrilateral ABCD, AD > AB > CD > BC; prove ∠ABC > ∠ADC. Which is the greater, ∠BAD or ∠BCD?
8. ABC is a triangle; the external bisector of ∠BAC cuts BC produced at D; prove (i) AB > AC; (ii) CD > AC.
9. ABC is a triangle; the bisector of ∠BAC cuts BC at D; if AB > AC, prove BD > DC.
10. ABC is an acute-angled triangle, such that ∠ABC = 2∠ACB; prove AC < 2AB.
11. ABCD is a quadrilateral; prove that AB + BC + CD > AD.
12. Prove that any side of a triangle is less than half its perimeter.
13. How many triangles can be drawn such that two of the sides are of lengths 4 feet, 7 feet, and such that the third side contains a whole number of feet?
14. ABC is a △; D is any point on BC; prove that AD < ½(AB + BC + CA).
15. ABCD is a quadrilateral; AB < BC; ∠BAD < ∠BCD; prove AD > CD.
16. ABC is a △; P is any point on BC; prove that AP is less than one of the lines AB, AC.
17. O is any point inside the triangle ABC; prove that (i) ∠BOC > ∠BAC; (ii) BO + OC < BA + AC.
18. A, B are any two points on the same side of CD, A' is the

image of **A** in **CD**; **A'B** cuts **CD** at **O**; **P** is any other point on **CD**; prove that **AP + PB > AO + OB**.
19. **AD** is a median of $\triangle$**ABC**; prove **AD** < $\frac{1}{2}$(**AB + AC**).
20. **O** is any point inside $\triangle$**ABC**; prove **OA + OB + OC** > $\frac{1}{2}$(**BC + CA + AB**).
21. In $\triangle$**ABC**, **BC > BA**; the perpendicular bisector **OP** of **AC** cuts **BC** at **P**; **Q** is any other point on **OP**; prove **AQ + QB > AP + PB**.
22. Prove that the sum of the diagonals of a quadrilateral is greater than the semiperimeter and less than the perimeter of the quadrilateral.

THE INTERCEPT THEOREM

Theorem 29

If **H, K** are the mid-points of the sides **AB, AC** of the triangle **ABC**, then (i) **HK** is parallel to **BC**.
(ii) **HK** = ½**BC**.

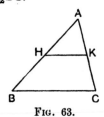

Fig. 63.

Theorem 30

If two lines **ABCDE, PQRST** are cut by the parallel lines **BQ, CR, DS** so that **BC** = **CD**, then **QR** = **RS**.

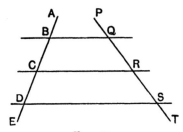

Fig. 64.

THE INTERCEPT THEOREM

EXERCISE XI

1. ABC is a △; H, K are the mid-points of AB, AC; P is any point on BC; prove HK bisects AP.
2. In △ABC, ∠BAC = 90°; D is the mid-point of BC; prove that AD = ½BC. [From D, drop a perpendicular to AC.]
3. In Fig. 65, if AC = CB and if AP, BQ, CR are parallel, prove that CR = ½(AP + BQ).

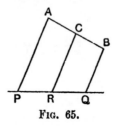

FIG. 65.

4. In Fig. 66, if AC = CB, and if AP, BQ, CR are parallel, prove that CR = ½(BQ − AP).

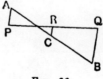

FIG. 66.

5. P, Q, R, S are the mid-points of the sides AB, BC, CD, DA of the quadrilateral ABCD; prove that PQ is equal and parallel to SR.
6. In △ABC, ∠ABC = 90°; BCX is an equilateral triangle; prove that the line from X parallel to AB bisects AC.
7. ABC is a △; H, K are the mid-points of AB, AC; BK, CH are produced to X, Y so that BK = KX and CH = HY; prove that XY = 2BC.

THE INTERCEPT THEOREM

8. O is a fixed point; P is a variable point on a fixed line AB; find the locus of the mid-point of OP.

9. O is a fixed point; P is a variable point on a fixed circle, centre A; prove that the locus of the mid-point of OP is a circle whose centre is at the mid-point of OA.

10. Prove that the lines joining the mid points of opposite sides of any quadrilateral bisect each other.

11. If the diagonals of a quadrilateral are equal and cut at right angles, prove that the mid-points of the four sides are the corners of a square.

12. ABCD is a quadrilateral; if AB is parallel to CD, prove that the mid-points of AD, BC, AC, BD lie on a straight line.

13. ABC is a $\triangle$; AX, AY are the perpendiculars from A to the bisectors of the angles ABC, ACB: prove that XY is parallel to BC.

14. ABCD is a quadrilateral such that BD bisects $\angle$ ABC and $\angle$ ADB $= 90° = \angle$ BCD; AH is the perpendicular from A to BC; prove AH $=$ 2CD.

15. AD, BE are altitudes of $\triangle$ABC and intersect at H; P, Q, R are the mid-points of HA, AB, BC; prove that $\angle$ PQR $= 90°$.

16. ABCD is a quadrilateral, having AB parallel to CD; P, Q, R, S are the mid-points of AD, BD, AC, BC; prove that (i) PQ $=$ RS; (ii) PS $= \frac{1}{2}$(AB + CD); (iii) QR $= \frac{1}{2}$(AB$\sim$CD).

17. ABC is a $\triangle$; D is the mid-point of BC; P is the foot of the perpendicular from B to the bisector of $\angle$ BAC; prove that DP $= \frac{1}{2}$(AB$\sim$AC).

18. ABC is a $\triangle$; D is the mid-point of BC; Q is the foot of the perpendicular from B to the external bisector of $\angle$ BAC; prove that DQ $= \frac{1}{2}$(AB + AC).

19. ABCD is a quadrilateral having AB $=$ CD; P, Q, R, S are the mid-points of AD, AC, BD, BC; prove that PS is perpendicular to QR.

20. In Fig. 67, if BD = DC and AP = AQ, prove that BP = CQ and AP = ½(AB + AC).

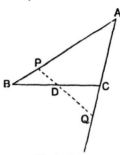

Fig. 67.

21. A square box ABCD, each edge 13″, rests in the rack of a railway carriage and against the wall: the point of contact E, is 1½″ from the wall; CE = ED. Prove that C is 5″ from the wall, and find the distances of A, D from the wall.

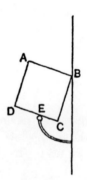

Fig. 68.

22. ABC is a △; E, F are the mid-points of AC AB; BE cuts CF at G; AG is produced to X so that AG = GX and cuts BC at D; prove that (i) GBXC is a parallelogram; (ii) DG = ½GA = ⅓DA.
23. ABCD is a parallelogram; XY is any line outside it; AP, BQ, CR, DS are perpendiculars from A, B, C, D to XY; prove that AP + CR = BQ + DS.
24.* The diagonals AC, BD of the square ABCD intersect at O;

THE INTERCEPT THEOREM

the bisector of $\angle$ BAC cuts BO at X, BC at Y; prove that CY = 2OX.

25*. Two equal circles, centres A, B, intersect at O; a third equal circle passes through O and cuts the former circles at C, D; prove that AB is equal and parallel to CD.

26*. A, B are fixed points; P is a variable point; PAST, PBXY are squares outside $\triangle$PAB; prove that the mid-point of SX is fixed. [Drop perpendiculars from S, X to AB.]

27*. ABCD is a quadrilateral having AD = BC; E, F are the mid-points of AB, CD; prove that EF is equally inclined to AD and BC. [Complete the parallelogram DABH: bisect at K; join BK, KF.]

RIDERS ON BOOK III

SYMMETRICAL PROPERTIES OF A CIRCLE

THEOREM 31

AB is a chord of a circle, centre **O**.
 (1) If **N** is the mid-point of **AB**, then $\angle \mathbf{ONA} = 90°$.
 (2) If **ON** is the perpendicular from **O** to **AB**, then **AN = NB**.

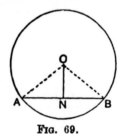

FIG. 69.

THEOREM 32

AB, CD are chords of a circle, centre **O**.
 (1) If **AB = CD**, then **AB** and **CD** are equidistant from **O**.
 (2) If **AB** and **CD** are equidistant from **O**, then **AB = CD**.

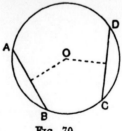

FIG. 70.

A corresponding property holds for equal circles.

SYMMETRICAL PROPERTIES OF A CIRCLE

EXERCISE XII

1. **AB** is a chord of a circle of radius 10 cms.; **AB** = 8 cms.; find the distance of the centre of the circle from **AB**.

2. A chord of length 10 cms. is at a distance of 12 cms. from the centre of the circle; find the radius.

3. A chord of a circle of radius 7 cms. is at a distance of 4 cms. from the centre; find its length.

4. **ABC** is a $\triangle$ inscribed in a circle; **AB** = **AC** = 13″, **BC** = 10″; calculate the radius of the circle.

5. In a circle of radius 5 cms., there are two parallel chords of lengths 4 cms., 6 cms.; find the distance between them.

6. Two parallel chords **AB**, **CD** of a circle are 3″ apart; **AB** = 4″, **CD** = 10″; calculate the radius of the circle.

7. An equilateral triangle, each side of which is 6 cms., is inscribed in a circle; find its radius.

8. The perpendicular bisector of a chord **AB** cuts **AB** at **C** and the circle at **D**; **AB** = 6″, **CD** = 1″; calculate the radius of the circle.

9. **ABC** is a straight line, such that **AB** = 1″, **BC** = 4″; **PBQ** is the chord of the circle on **AC** as diameter, perpendicular to **AC**; find **PQ**.

10. **P** is a point on the diameter **AB** of a circle; **AP** = 2″, **PB** = 8″; find the length of the shortest chord which passes through **P**.

11. The centres of two circles of radii 3″, 4″ are at a distance 5″ apart; find the length of their common chord.

12. Two concentric circles are of radii 3″, 5″; a line **PQRS** cuts one at **P**, **S** and the other at **Q**, **R**; if **QR** = 2″, find **PQ**.

13. A variable line **PQRS** cuts two fixed concentric circles of radii a'', b'' at **P**, **S** and **Q**, **R**; if **PQ** = x'', **QR** = y'', find an equation between x, y, a, b, and prove that **PQ** . **QS** is constant.

14. A crescent is formed of two circular arcs of equal radius (see Fig. 71); the perpendicular bisector of AB cuts the crescent at C, D; if CD = 3 cms., AB = 10 cms., find the radii.

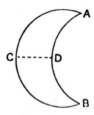

Fig. 71.

15. In Fig. 72, ABCD is the section of a lens; AB = CD = x; BP = PC = y; PQ = z; AB, QP, DC are perpendicular to BC; calculate in terms of x, y, z the radius of the circular arc AQD.

Fig. 72.

16. AB is a chord of a circle, centre O; T is any point equidistant from A and B; prove OT bisects ∠ATB.

17. Two circles, centres A, B, intersect at X, Y; prove that AB bisects XY at right angles.

18. Two circles, centres A, B, intersect at C, D; PCQ is a line parallel to AB cutting the circles at P, Q; prove PQ = 2AB.

19. Two circles, centres A, B, intersect at X, Y; PQ is a chord of one circle, parallel to XY; prove AB bisects PQ.

20. A line PQRS cuts two concentric circles at P, S and Q, R; prove PQ = RS.

21. ABC is a triangle inscribed in a circle; if ∠BAC = 90°, prove that the mid-point of BC is the centre of the circle.

SYMMETRICAL PROPERTIES OF A CIRCLE 59

22. In Fig. 73, if **PQ** is parallel to **RS**, prove **PQ** = **RS**.

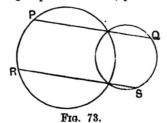

Fig. 73.

23. **APB, CPD** are intersecting chords of a circle, centre **O**; if **OP** bisects ∠ **APC**, prove **AB** = **CD**.
24. The diagonals of the quadrilateral **ABCD** meet at **O**; circles are drawn through **A, O, B ; B, O, C ; C, O, D ; D, O, A**; prove that their four centres are the corners of a parallelogram.
25. **AOB, COD** are two intersecting chords of a circle; if **AB** = **CD**, prove **AO** = **CO**.
26. In Fig. 74, **A, C, B** are the centres of three unequal circles; if **AC** = **CB**, prove **PQ** = **RS**.

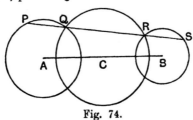

Fig. 74.

27. **AB, CD** are two chords of a circle, centre **O**; if **AB** > **CD**, prove **O** is nearer to **AB** than to **CD**.
28. Two circles, centres **A, B**, intersect at **C, D**; **PCQ** is a line cutting the circles at **P, Q**; prove **PQ** is greatest when it is parallel to **AB**.
29*. **P** is any point on a diameter **AB** of a circle; **QPR** is a chord such that ∠ **APQ** = 45°; prove that $AB^2 = 2PQ^2 + 2PR^2$.
30*. **ABC** is a △ inscribed in a circle, centre **O**; **X, Y, Z** are the images of **O** in **BC, CA, AB**; prove that **AX, BY, CZ** bisect each other.
31*. **AB, CD** are two perpendicular chords of a circle, centre **O**; prove that $AC^2 + BD^2 = 4OA^2$.

ANGLE PROPERTIES OF A CIRCLE (1)

Theorem 33

If **AB** is an arc of a circle, centre **O**, and if **P** is any point on the remaining part of the circumference, then the angle which arc **AB** subtends at **O** equals $2 \angle APB$,
$$\angle AOB = 2 \angle APB.$$

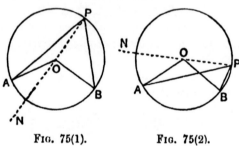

Fig. 75(1).   Fig. 75(2).

Theorem 34

(1) If **APQB** is a circle, $\angle APB = \angle AQB$.
(2) If **AB** is a diameter of a circle **APB**, $\angle APB = 90°$.

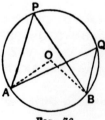 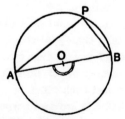

Fig. 76.   Fig. 77.

ANGLE PROPERTIES OF A CIRCLE (1)

Theorem 35

(1) If **ABCD** is a cyclic quadrilateral, $\angle ABC + \angle ADC = 180°$.
(2) If the side **AD** of the cyclic quadrilateral **ABCD** is produced to **P**,

$$\angle PDC = \angle ABC.$$

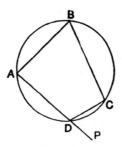

Fig. 78.

ANGLE PROPERTIES OF A CIRCLE (1)
EXERCISE XIII

1. **ABC** is a △ inscribed in a circle, centre **O**; $\angle$**AOC** = 130° $\angle$**BOC** = 150°, find $\angle$**ACB**.
2. **AB, CD** are perpendicular chords of a circle; $\angle$**BAC** = 35°, find $\angle$**ABD**.
3. **ABCD** is a quadrilateral such that **AB** = **AC** = **AD**; if $\angle$**BAD** = 140°, find $\angle$**BCD**.
4. **ABCD** is a quadrilateral inscribed in a circle; **AB** is a diameter; $\angle$**ADC** = 127°; find $\angle$**BAC**.
5. Two chords **AB, CD** when produced meet at **O**; $\angle$**OAD** = 31°; $\angle$**AOC** = 42°; find $\angle$**OBC**.
6. Two circles **APRB, AQSB** intersect at **A, B**; **PAQ, RBS** are straight lines; if $\angle$**QPR** = 80°, $\angle$**PRS** = 70°, find $\angle$**PQS**, $\angle$**QSR**.
7. **P, Q, R** are points of a circle, centre **O**; $\angle$**POQ** = 54°, $\angle$**OQR** = 36°; **P, R** are on opposite sides of **OQ**; find $\angle$**QPR** and $\angle$**PQR**.
8. The diagonals of the cyclic quadrilateral **ABCD** meet at **O**; $\angle$**BAC** = 42°, $\angle$**BOC** = 114°, $\angle$**ADB** = 33°; find $\angle$**BCD**.
9. **ABCD** is a cyclic quadrilateral, **EABF** is a straight line; $\angle$**EAD** = 82°, $\angle$**FBC** = 74°, $\angle$**BDC** = 50°; find angle between **AC, BD**.
10. Two chords **AB, DC** of a circle, centre **O**, are produced to meet at **E**; $\angle$**AOB** = 100°, $\angle$**EBC** = 72°, $\angle$**ECB** = 84°; find $\angle$**COD**.
11. (i) In Fig. 79, if $y = 32°$, $z = 40°$, find x.
    (ii) If $y + z = 90°$, prove $x = 45°$.

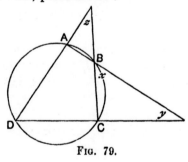

Fig. 79.

ANGLE PROPERTIES OF A CIRCLE (1)

12. D is a point on the base BC of $\triangle$ABC; H, K are the centres of the circles ADB, ADC; if $\angle$AHD $= 70°$, $\angle$AKD $= 80°$, find $\angle$BAC.
13. In Fig. 79, if AC cuts BD at O, if $y = 20°$, $z = 40°$, $\angle$BOC $= 100°$, prove $\angle$BAC $= 2\angle$BCA.

---

14. AB, XY are parallel chords of a circle; AY cuts BX at O; prove OX $=$ OY.
15. Two circles BAPR, BASQ cut at A, B; PAQ, RAS are straight lines; prove $\angle$PBR $= \angle$QBS.
16. AB is a chord of a circle, centre O; P is any point on the minor arc AB; prove $\angle$AOB $+ 2\angle$APB $= 360°$.
17. ABCD is a cyclic quadrilateral; if AC bisects the angles at A and C, prove $\angle$ABC $= 90°$.
18. Two lines OAB, OCD cut a circle at A, B and C, D; prove $\angle$OAD $= \angle$OCB.
19. AB is a diameter of a circle APQRB; prove $\angle$APQ $+ \angle$QRB $= 270°$.
20. ABCDEF is a hexagon inscribed in a circle; prove that $\angle$FAB $+ \angle$BCD $+ \angle$DEF $= 360°$.
21. Two circles ABPR, ABQS cut at A, B; PBQ, RAS are straight lines; prove PR is parallel to QS.
22. A, B, C are three points on a circle, centre O; prove that $\angle$BAC $= \angle$OBA $\pm \angle$OCA.
23. A, B, C, P are four points on a circle; prove that a triangle whose sides are parallel to PA, PB, PC is equiangular to $\triangle$ABC.
24. AP, AQ are diameters of the circles APB, AQB; prove that PBQ is a straight line.
25. OA is a radius of a circle, centre O; AP is any chord; prove that the circle on OA as diameter bisects AP.
26. Two chords AOB, COD of a circle intersect at O; if AO $=$ AC, prove DO $=$ BD.
27. APC is an arc, less than a semicircle, of a circle, centre O; AQOC is another circular arc; prove $\angle$APC $= \angle$PAQ $+ \angle$PCQ.

28. ABC is a △ inscribed in a circle, centre O; D is the midpoint of BC; prove ∠BOD = ∠BAC.
29. OA, OB, OC are three equal lines; if ∠AOB = 90°, prove ∠ACB = 45° or 135°.
30. Two lines OAB, OCD cut a circle at A, B and C, D; if OB = BD, prove OC = CA.
31. ABCD is a rectangle; any circle through A cuts AB, AC, AD at X, Y, Z; prove that ABD, XYZ are equiangular triangles.
32. In Fig. 80, O is the centre of the circle; prove ∠AOC + ∠BOD = 2∠AEC.

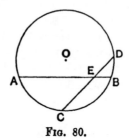

Fig. 80.

33. ABCD is a cyclic quadrilateral; AD, BC are produced to meet at E; AB, DC are produced to meet at F; the circles EDC, FBC cut at X; prove EXF is a straight line.
34. AB, CD are perpendicular chords of a circle, centre O; prove ∠DAB = ∠OAC.
35. In △ABC, AB = AC; ABD is an equilateral triangle; prove that ∠BCD = 30° or 150°.
36. ABC is a △; D is a point on BC; H, K are the centres of the circles ADB, ADC; if H, D, K, A are concyclic, prove ∠BAC = 90°.
37. ABC is a △; the bisectors of ∠s ABC, ACB intersect at I and meet AC, AB at P, Q; if A, Q, I, P are concyclic, prove ∠BAC = 60°.
38. Two lines EBA, ECD cut a circle ABCD at B, A and C, D; O is the centre; prove ∠AOD − ∠BOC = 2∠BEC.
39. ACB, ADB are two arcs on the same side of AB; a straight line ACD cuts them at C, D; if the centre of the circle ADB lies on the arc ACB, prove CB = CD.

ANGLE PROPERTIES OF A CIRCLE (1)

40. **ABCD** is a quadrilateral inscribed in a circle; **BA, CD** when produced meet at **E**; **O** is the centre of the circle **EAC**; prove that **BD** is perpendicular to **OE**.
41. **ABC** is a △ inscribed in a circle; **AOX, BOY, COZ** are three chords intersecting at a point **O** inside △**ABC**; prove ∠**YXZ** = ∠**BOC** − ∠**BAC**.
42. **D** is any point on the side **AB** of △**ABC**; points **E, F** are taken on **AC, BC** so that ∠**EDA** = 60° = ∠**FDB**; a circle is drawn through **D, E, F** and cuts **AB** again at **G**; prove △**EFG** is equilateral.
43. **ABC** is a △; a line **PQR** cuts **BC** produced, **CA**, **AB** at **P, Q, R**; if **B, C, Q, R** are concyclic, prove the bisectors of ∠s **BPR, BAC** are at right angles.
44. **APXBYQ** is a circle; **AB** bisects ∠**PAQ** and ∠**XAY**; prove **PQ** is parallel to **XY**.
45. **ABC** is a △; the bisectors of ∠s **ABC, ACB** meet at **I**; the circle **BIC** cuts **AB, AC** again at **P, Q**; prove **AB** = **AQ** and **AC** = **AP**.
46. **AB** is a diameter of a circle **AQBR**; **AQ, BR** meet when produced at **O**; use an area formula to prove that **BQ . AO** = **AR . BO**.
47. **ABC** is a △; the bisectors of ∠s **ABC, ACB** intersect at **I**, and cut **AC, AB** at **Y, Z**; the circles **BIZ, CIY** meet again at **X**; prove ∠**YXZ** + ∠**BIC** = 180°.
48. **ABC** is a triangle inscribed in a circle; **AB** = **AC**; **BC** is produced to **D**; **AD** cuts the circle at **E**; prove ∠**ACE** = ∠**ADB**.
49*. **AOB, COD** are perpendicular chords of a circle **ACBD**; prove that the perpendicular from **O** to **AD** bisects, when produced, **BC**.
50*. **ABCD** is a quadrilateral inscribed in a circle, centre **O**; if **AC** is perpendicular to **BD**, prove that the perpendicular from **O** to **AD** equals $\frac{1}{2}$**BC**.
51*. **OC** is a radius perpendicular to a diameter **AOB** of a circle; **P, Q** are the feet of the perpendiculars from **A, B** to any line through **C**; prove that **PC** = **QB** and that $AP^2 + BQ^2 = 2OC^2$.
52*. Two given circles **ABP, ABQ** intersect at **A, B**; a variable

line **PAQ** meets them at **P, Q**; prove ∠ **PBQ** is of constant size.

53*. **ABC** is a given △; **P** is a variable point on a given circle which passes through **B, C**; if **P, A** are on the same side of **BC**, prove ∠ **PBA** − ∠ **PCA** is constant.

54*. In Fig. 81, the circles are given; prove ∠ **PRQ** is of constant size.

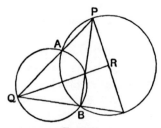

Fig. 81.

55*. **AB** is a fixed chord, and **AP** a variable chord of a given circle; **C, Q** are the mid-points of **AB, AP**; prove ∠ **AQC** has one of two constant values.

56*. A variable circle passes through a fixed point **A** and cuts two given parallel lines at **P, Q** such that ∠ **PAQ** = 90°; prove that the circle passes through a second fixed point.

57*. Two circles **PRQ, PSQ** intersect at **P, Q**; the centre **O** of circle **PRQ** lies on circle **PSQ**; the diameter **PS** of circle **PSQ** cuts circle **PRQ** at **R**; prove **QR** is parallel to **OP**.

ANGLE PROPERTIES OF A CIRCLE (2)

THEOREM 40

If **P** is any point on a circle, centre **O**, and if **PX** is the tangent at **P**, then $\angle \mathbf{OPX} = 90°$.

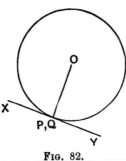

FIG. 82.

THEOREM 41

If **PA** is any chord of a circle **PKA**, and if **PX** is the tangent at **P**, **K** and **X** being on opposite sides of **PA**, then $\angle \mathbf{APX} = \angle \mathbf{AKP}$.

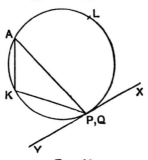

FIG. 83.

ANGLE PROPERTIES OF A CIRCLE (2)

EXERCISE XIV

1. A line **TBC** cuts a circle **ABC** at **B, C**; **TA** is a tangent; if $\angle$**TAC** $= 118°$, $\angle$**ATC** $= 26°$, find $\angle$**ABC**.
2. **ABC** is a minor arc of a circle; the tangents at **A, C** meet at **T**; if $\angle$**ATC** $= 54°$, find $\angle$**ABC**.
3. **AOC, BOD** are chords of a circle **ABCD**; the tangent at **A** meets **DB** produced at **T**; if $\angle$**ATD** $= 24°$, $\angle$**COD** $= 82°$, $\angle$**TBC** $= 146°$, find $\angle$**BAC**. Find also the angle between **BD** and the tangent at **C**.
4. The sides **BC, CA, AB** of a $\triangle$ touch a circle at **X, Y, Z**; $\angle$**ABC** $= 64°$, $\angle$**ACB** $= 52°$; find $\angle$**XYZ**, $\angle$**XZY**.
5. Three of the angles of a quadrilateral circumscribing a circle are $70°$, $84°$, $96°$ in order; find the angles of the quadrilateral whose vertices are the points of contact.
6. **TBP, TCQ** are tangents to the circle **ABC**; $\angle$**PBA** $= 146°$, $\angle$**QCA** $= 128°$; find $\angle$**BAC** and $\angle$**BTC**.
7. In $\triangle$**ABC**, $\angle$**ABC** $= 50°$, $\angle$**ACB** $= 70°$; a circle touches **BC**, **AC** produced, **AB** produced at **X, Y, Z**; find $\angle$**YXZ**.

---

8. A chord **AB** of a circle is produced to **T**; **TC** is a tangent from **T** to the circle; prove $\angle$**TBC** $= \angle$**ACT**.
9. Two circles **APB, AQB** intersect at **A, B**; **AP, AQ** are the tangents at **A**, prove $\angle$**ABP** $= \angle$**ABQ**.
10. **DA** is the tangent at **A** to the circle **ABC**; if **DB** is parallel to **AC**, prove $\angle$**ADB** $= \angle$**ABC**.
11. In $\triangle$**ABC**, **AB** $=$ **AC**; **D** is the mid-point of **BC**; prove that the tangent at **D** to the circle **ADC** is perpendicular to **AB**.
12. **BC, AD** are parallel chords of the circle **ABCD**; the tangent at **A** cuts **CB** produced at **P**; **PD** cuts the circle at **Q**; prove $\angle$**PAQ** $= \angle$**BPQ**.
13. Two circles **ACB, ADB** intersect at **A, B**; **CA, DB** are tangents to circles **ADB, ACB** at **A, B**; prove **AD** is parallel to **BC**.
14. **CA, CB** are equal chords of a circle; the tangent **ADE** at **A** meets **BC** produced at **D**; prove $\angle$**BDE** $= 3\angle$**CAD**.
15. The bisector of $\angle$**BAC** meets **BC** at **D**; a circle is drawn

touching BC at D and passing through A; if it cuts AB, AC at P, Q, prove $\angle$ PDB = $\angle$ QDC.

16. Two circles APB, AQB intersect at A, B; AQ, AP are the tangents at A; if PBQ is a straight line, prove $\angle$ PAQ = $90°$.

17. ABCD is a quadrilateral inscribed in a circle; the tangents at A, C meet at T; prove $\angle$ ATC = $\angle$ ABC $\sim$ $\angle$ ADC.

18. Two circles intersect at A, B; the tangents at B meet the circles at P, Q; if $\angle$ PBQ is acute, prove $\angle$ PAQ = $2 \angle$ PBQ. What happens if $\angle$ PBQ is obtuse?

19. ABC is a $\triangle$ inscribed in a circle; the tangent at C meets AB produced at T; the bisector of $\angle$ ACB cuts AB at D; prove TC = TD.

20. AOB is a diameter of a circle, centre O; the tangent at B meets any chord AP at T; prove $\angle$ ATB = $\angle$ OPB.

21. ABCDE is a pentagon inscribed in a circle; AT is the tangent at A, T and D being on opposite sides of AB; prove $\angle$ BCD + $\angle$ AED = $180° + \angle$ BAT.

22. In $\triangle$ABC, AB = AC; a circle is drawn to touch BC at B and to pass through A; if it cuts AC at D, prove BC = BD.

23. In $\triangle$ABC, $\angle$ BAC = $90°$; D is any point on BC; DP, DQ are tangents at D to the circles ABD, ACD; prove $\angle$ PDQ = $90°$.

24. AB is a diameter of a circle ABC; TC is the tangent from a point T on AB produced; TD is drawn perpendicular to TA and meets AC produced at D; prove TC = TD.

25. EAF, CBD are tangents at the extremities of a chord AB of a circle, E and C being on the same side of AB; if AB bisects $\angle$ CAD, prove $\angle$ EAC = $\angle$ ADC.

26. Two circles touch internally at A; the tangent at any point P on the inner cuts the outer at Q, R; AQ, AR cut the inner at H, K; prove $\triangle$s PQH, APK are equiangular.

27. PQ is a common tangent to two circles CDP, CDQ; prove that $\angle$ PCQ + $\angle$ PDQ = $180°$.

28. Two chords AOB, COD of a circle cut at O; the tangents at A, C meet at X; the tangents at B, D meet at Y; prove $\angle$ AXC + $\angle$ BYD = $2 \angle$ AOD.

29. I is the centre of a circle touching the sides of $\triangle$ABC; a larger concentric circle is drawn; prove that it cuts off equal portions from AB, BC, CA.

30. **PQ, PR** are equal chords of a circle; **PQ** and the tangent at **R** intersect at **T**; prove $\angle \mathbf{PRQ} = 60° \pm \tfrac{1}{3} \angle \mathbf{PTR}$.
31. The diameter **AB** of a circle, centre **O**, is produced to **T** so that **OB = BT**; **TP** is a tangent to the circle; prove **TP = PA**.
32. The bisector of $\angle$ **BAC** cuts **BC** at **D**; a circle is drawn through **D** and to touch **AC** at **A**; prove that its centre lies on the perpendicular from **D** to **AB**.
33. Three circles, centres **A, B, C**, have a common point of intersection **O**; also their common chords are equal; prove that **O** is the centre of the circle inscribed in $\triangle$**ABC**.
34. **AB** is a chord of a circle; the tangents at **A, B** meet at **T**; **AP** is drawn perpendicular to **AB**, and **TP** is drawn perpendicular to **TA**; prove that **PT** equals the radius.
35. Two circles **ABD, ACE** intersect at **A**; **BAC, DAE** are straight lines; prove that the angle between **DB** and **CE** equals the angle between the tangents at **A**.
36. Assuming the result of ex. 21 (page 63), what special cases can be obtained by taking (i) **Q** very close to **S**, (ii) **Q** very close to **B**, (iii) **A** very close to **B**?
37. **A, B** are given points on a given circle; **P** is a variable point on the circle; the circles whose diameters are **AB** and **AP** intersect at **Q**. Find the position of **Q** when **P** is very close to **B**.
38. **OA** is a chord of a circle, centre **C**; **T** is a point on the tangent at **O** such that **OA = OT** and $\angle$ **AOT** is acute; **TA** is produced to cut **OC** at **B**; prove that $\angle \mathbf{OBA} = \tfrac{1}{2} \angle \mathbf{OCA}$. Find the position of **B** when **A** is very close to **O**.

PROPERTIES OF EQUAL ARCS AND EQUAL CIRCLES

Theorem 37.

H, K are the centres of two equal circles **APB, CQD**.
  (i) If $\angle$ **AHB** = $\angle$ **CKD**, then arc **AB** = arc **CD**.
  (ii) If $\angle$ **APB** = $\angle$ **CQD**, then arc **AB** = arc **CD**.

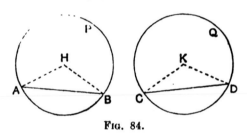

Fig. 84.

Theorem 38.

H, K are the centres of two equal circles **APB, CQD**.
If arc **AB** = arc **CD**, then (i) $\angle$ **AHB** = $\angle$ **CKD**,
  and (ii) $\angle$ **APB** = $\angle$ **CQD**.

Theorem 39.

APB, CQD are two equal circles.
  (i) If chord **AB** = chord **CD**, then arc **AB** = arc **CD**.
  (ii) If arc **AB** = arc **CD**, then chord **AB** = chord **CD**.
These properties also hold for equal arcs in the *same* circle.

PROPERTIES OF EQUAL ARCS AND EQUAL CIRCLES

EXERCISE XV

1. **ABCD** is a square and **AEF** is an equilateral triangle inscribed in the same circle; calculate the angles of $\triangle$**ECD**.
2. **AB** is a side of a regular hexagon and **AC** of a regular octagon inscribed in the same circle; calculate the angles of $\triangle$**ABC**.
3. **ABCD** is a quadrilateral inscribed in a circle; **AC** cuts **BD** at **O**: **DA**, **CB** when produced meet at **E**; **AB**, **DC** when produced meet at **F**; if $\angle$**AEB** $= 55°$, $\angle$**BFC** $= 35°$, $\angle$**DOC** $= 85°$, prove arc **BC** = twice arc **AB**.
4. **ABC** is a triangle inscribed in a circle: the tangent at **A** meets **BC** produced at **T**; $\angle$**BAT** $= 135°$, $\angle$**ATB** $= 30°$; find the ratio of the arcs **AB** and **AC**.
5. **A**, **B** are two points on the circle **ABCD** such that the minor arc **AB** is half the major arc **AB**; $\angle$**DAB** $= 74°$; arc **BC** = arc **CD**; calculate $\angle$**ABD** and $\angle$**BDC**.
6. **ABCD** is a quadrilateral inscribed in a circle; $\angle$**ADB** $= 25°$, $\angle$**DBC** $= 65°$; prove arc **AB** + arc **CD** = arc **BC** + arc **AD**.

---

7. **AB**, **CD** are parallel chords of a circle; prove arc **AD** = arc **BC**.
8. **ABCD** is a cyclic quadrilateral; if **AB** = **CD**, prove $\angle$**ABC** = $\angle$**BCD**.
9. A circle **AOBP** passes through the centre **O** of a circle **ABQ**; prove that **OP** bisects $\angle$**APB**.
10. **ABP**, **ABQ** are two equal circles; **PBQ** is a straight line; prove **AP** = **AQ**.
11. **AB**, **BC**, **CD** are equal chords of a circle, centre **O**; prove that **AC** cuts **BD** at an angle equal to $\angle$**AOB**.
12. **ABCD** is a square and **APQ** an equilateral triangle inscribed in the same circle, **P** being between **B** and **C**; prove arc **BP** $= \frac{1}{2}$ arc **PC**.
13. On a clock-face, prove that the line joining 4 and 7 is perpendicular to the line joining 5 and 12.

PROPERTIES OF EQUAL ARCS AND CIRCLES

14. **X, Y** are the mid-points of the arcs **AB, AC** of a circle; **XY** cuts **AB, AC** at **H, K**; prove **AH = AK**.
15. **APB, AQB** are two equal circles; **AP** is a tangent to the circle **AQB**, prove **AB = BP**.
16. **ABCD** is a rectangle inscribed in a circle; **DP** is a chord equal to **DC**; prove **PB = AD**.
17. A hexagon is inscribed in a circle; if two pairs of opposite sides are parallel, prove that the third pair is also parallel.
18. **ABC** is a $\triangle$ inscribed in a circle; any circle through **B, C** cuts **AB, AC** again at **P, Q**; **BQ, CP** are produced to meet the circle **ABC** at **R, S**; prove **AR = AS**.
19. **ABCDEF** is a hexagon inscribed in a circle; if $\angle$ **ABC** = $\angle$ **DEF**, prove **AF** is parallel to **CD**.
20. **CD** is a quadrant of the circle **ACDB**; **AB** is a diameter; **AD** cuts **BC** at **P**; prove **AC = CP**.
21. **ABC** is a $\triangle$ inscribed in a circle, centre **O**; **P** is any point on the side **BC**; prove that the circles **OBP, OCP** are equal.
22. In $\triangle$**ABC**, **AB = AC**; **BC** is produced to **D**; prove that the circles **ABD, ACD** are equal.
23. **ABCD** is a quadrilateral inscribed in a circle; **CD** is produced to **F**; the bisector of $\angle$ **ABC** cuts the circle at **E**; prove that **DE** bisects $\angle$ **ADF**.
24. **ABCD** is a cyclic quadrilateral; **BC** and **AD** are produced to meet at **E**; a circle is drawn through **A, C, E** and cuts **AB, CD** again at **P, Q**; prove **PE = EQ**.
25. **AB, AC** are equal chords of a circle; **BC** is produced to **D** so that **CD = CA**; **DA** cuts the circle at **E**; prove that **BE** bisects $\angle$ **ABC**.
26. **ABC** is an equilateral triangle inscribed in a circle; **H, K** are the mid-points of the arcs **AB, AC**; prove that **HK** is trisected by **AB, AC**.
27. **AB, BC** are two chords of a circle (**AB > BC**); the minor arc **AB** is folded over about the chord **AB** and cuts **AC** at **D**; prove **BD = BC**.
28. **ABCD** is a quadrilateral inscribed in a circle; **X, Y, Z, W** are the mid-points of the arcs **AB, BC, CD, DA**; prove that **XZ** is perpendicular to **YW**.
29. In $\triangle$**ABC**, **AB > AC**; the bisectors of $\angle$s **ABC, ACB** meet

at I; the circle BIC cuts AB, AC at P, Q; prove PI = IC and QI = IB.

30. ABC is a triangle inscribed in a circle, centre O; PQ is the diameter perpendicular to BC, P and A being on the same side of BC; prove ∠ABC ∼ ∠ACB = ∠POA.

31*. In Fig. 85, the circles are equal and AD = BC; prove XBYD is a parallelogram.

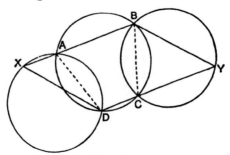

Fig. 85.

32*. In △ABC, AB = AC; D is any point on BC; X, Y are the centres of the circles ABD, ACD; XP, YQ are the perpendiculars to AB, AC; prove XP = YQ.

33*. AB, CD are two perpendicular chords of a circle, centre O; prove that $AC^2 + BD^2 = 4OA^2$. [Use Theorem 25(2).]

34*. $A_1 A_2 A_3 \ldots A_{2n}$ is a regular polygon of $2n$ sides; if $2n > p > q > r > s$, prove that $A_p A_r$ is perpendicular to $A_q A_s$ if $p + r = q + s + n$.

35*. ABC is an equilateral triangle inscribed in a circle; D, E are points on the arcs AB, BC such that AD = BE, prove AD + DB = AE.

36*. C is the mid-point of a chord AB of a circle; D, E are points on the circle on opposite sides of AB such that ∠DAC = ∠AEC; prove that ∠ADC = ∠EAC.

37*. P, Q, R are points on the sides BC, CA, AB of △ABC such that ∠PQR = ∠ABC and ∠PRQ = ∠ACB; prove that the circles AQR, BRP, CPQ meet at a common point, K say, and are equal; prove also that (i) ∠BKC = 2∠BAC; (ii) AK = BK = CK; (iii) PK is perpendicular to QR.

38*. Two fixed circles cut at A, B; P is a variable point on one;

PA, PB when produced cut the other at QR; prove QR is of constant length.

39*. A is a fixed point on a fixed circle; B is a fixed point on a fixed line BC; a variable circle through A, B cuts BC at P and the fixed circle at Q; prove that PQ cuts the fixed circle at a fixed point.

LENGTHS OF TANGENTS AND CONTACT OF CIRCLES

THEOREM 42

If **TP, TQ** are the tangents from **T** to a circle, centre **O**, then (i)   **TP = TQ**.
(ii) ∠ **TOP = TOQ**.
(iii) **OT** bisects ∠ **PTQ**.

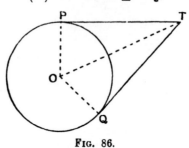

FIG. 86.

THEOREM 43

If two circles, centres **A, B**, touch, internally or externally, at **P**, then **APB** is a straight line.

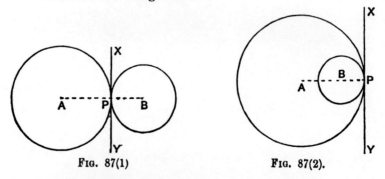

FIG. 87(1).   FIG. 87(2).

If the circles touch externally (Fig. 87(1)), the distance between the centres **AB** = *sum* of radii.

If the circles touch internally (Fig. 87(2)), the distance between the centres **AB** = *difference* of radii.

LENGTHS OF TANGENTS AND CONTACT OF CIRCLES

EXERCISE XVI

1. A circle, radius 5 cms., touches two concentric circles and encloses the smaller: the radius of the larger circle is 7 cms.: what is the radius of the smaller?
2. Three circles, centres **A, B, C**, touch each other externally; **AB** = 4″, **BC** = 6″, **CA** = 7″; find their radii.
3. In △**ABC**, **AB** = 4″, **BC** = 7″, **CA** = 5″; two circles with **B, C** as centres touch each other externally; a circle with **A** as centre touches the others internally; find their radii.
4. Fig. 88 is formed of three circular arcs of radii 6·7 cms., 2·2 cms., 3·1 cms.; **X, Y, Z** are the centres of the circles; find the lengths of the sides of △**XYZ**.

FIG. 88.

5. In Fig. 89, **AB** is a quadrant touching **AD** at **A** and the quadrant **BC** at **B**; ∠**ADC** = 90°, **AD** = 12″, **DC** = 9″; find the radii of the circles.

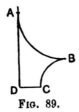

FIG. 89.

6. The distance between the centres of two circles of radii 4 cms., 7 cms. is 15 cms.; what is the radius of the least circle that can be drawn to touch them and enclose the smaller circle?

7. C is a point on AB such that AC = 5″, CB = 3″; calculate the radius of the circle which touches AB at C and also touches the circle on AB as diameter.
8. A, B are the centres of two circles of radii 5 cms., 3 cms.; AB = 12 cms.; BC is a radius perpendicular to BA; find the radius of a circle which touches the larger circle and touches the smaller circle at C. [Two answers.]
9. AB, BC are two equal quadrants touching at B; their radii are 12 cms.; find the radius of the circle which touches arc AB, arc BC, AC.

Fig. 90.

10. In △ABC, AB = 4″, BC = 6″, CA = 7″; a circle touches BC, CA, AB at X, Y, Z; find BX and AY.
11. In △ABC, AB = 3″, BC = 7″, CA = 9″; a circle touches CA produced, CB produced, AB at Q, P, R; find AQ, BR.
12. Two circles of radii 3 cms., 12 cms. touch each other externally; find the length of their common tangent.
13. The distance between the centres of two circles of radii 11 cms., 5 cms. is 20 cms.; find the lengths of their exterior and interior common tangents.
14. The distance between the centres of two circles is 10 cms., and the lengths of their exterior and interior common tangents are 8 cms., 6 cms.; find their radii.
15. ABCD is a square of side 7″; C is the centre of a circle of radius 3″; find the radius of the circle which touches this circle and touches AB at A.
16. In one corner of a square frame, side 3′, is placed a disc of radius 1′ touching both sides; find the radius of the largest disc which will fit into the opposite corner.
17. a, b are the lengths of the diameters of two circles which touch each other externally; t is the length of their common tangent; prove that $t^2 = ab$.
18. Two circles of radii 4 cms., 9 cms. touch each other externally;

find the radius of the circle which touches each of these circles and also their common tangent. [Two answers: use ex. 17.]

19. $OA = a''$, $OB = b''$, $\angle AOB = 90°$; two variable circles are drawn touching each other externally, one of them touches OA at A, and the other touches OB at B; if their radii are x'', y'', prove that $(x+a)(y+b)$ is constant. If $a = 8$, $b = 6$, $x = 4$, calculate y.

20. Four equal spheres, each of radius 1'', are fixed in contact with each other on a horizontal table, with their centres at the corners of a square; a fifth equal sphere rests on them; find the height of its centre above the table.

21. A circle touches the sides of △ABC at X, Y, Z; if Y, Z are the mid-points of AB, AC, prove that X is the mid-point of BC.

22. Two circles touch each other at A; any line through A cuts the circles at P, Q; prove that the tangents at P, Q are parallel.

23. ABCD is a quadrilateral circumscribing a circle, prove that AB + CD = BC + AD.

24. ABCD is a parallelogram; if the circles on AB and CD as diameters touch each other, prove that ABCD is a rhombus.

25. Two circles touch externally at A; PQ is their common tangent; prove that the tangent at A bisects PQ and that $\angle PAQ = 90°$.

26. In Fig. 91, prove AB − CD = BC − AD.

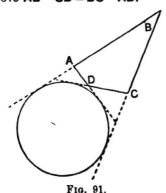

Fig. 91.

27. **ABCDEF** is a hexagon circumscribing a circle; prove that
    AB + CD + EF = BC + DE + FA.
28. In △**ABC**, ∠**BAC** = 90°; **O** is the mid-point of **BC**; circles are drawn with **AB** and **AC** as diameters; prove that two circles can be drawn with **O** as centre to touch each of these circles.
29. Two circles touch externally at **A**; **AB** is a diameter of one; **BP** is a tangent to the other; prove that ∠**APB** = 45° − ½∠**ABP**.
30. **ABCD** is a quadrilateral circumscribing a circle, centre **O**; prove ∠**AOB** + ∠**COD** = 180°.
31. Two circles touch internally at **A**; a chord **PQ** of one touches the other at **R**; prove ∠**PAR** = ∠**QAR**.
32. Two circles touch internally at **A**; any line **PQRS** cuts one at **P**, **S** and the other at **Q**, **R**; prove ∠**PAQ** = ∠**RAS**.
33. Two equal circles, centres **X**, **Y**, touch at **A**; **P**, **Q** are points, one on each circle such that ∠**PAQ** = 90°; prove that **PQ** is parallel to **XY**.
34. Two circles touching internally at **A**; **P**, **Q** are points, one on each circle, such that ∠**PAQ** = 90°; prove that the tangents at **P** and **Q** are parallel.
35. Two circles touch at **A**; any line **PAQ** cuts one circle at **P**, and the other at **Q**; prove that the tangent at **P** is perpendicular to the diameter through **Q**.
36. In △**ABC**, ∠**ABC** = 90°; a circle, centre **X**, is drawn to touch **AB** produced, **AC** produced, and **BC**; prove ∠**AXC** = 45°.
37. Two circles touch externally at **A**; a tangent to one of them at **P** cuts the other circle at **Q**, **R**; prove ∠**PAQ** + ∠**PAR** = 180°.
38. Two circles, centres **A**, **B**, touch externally at **P**; a third circle, centre **C**, encloses both, touching the first at **Q** and the second at **R**; prove ∠**BAC** = 2∠**PRQ**.
39. A circle, centre **A**, touches externally two circles, centres **B**, **C** at **X**, **Y**; **XY** cuts the circle, centre **C**, at **Z**; prove **BX** is parallel to **CZ**.
40. **PR**, **QR** are two circular arcs touching each other at **R**, and

touching the unequal lines OP, OQ at P, Q; prove that ∠PRQ = 180° − ½ ∠POQ (see Fig. 92).

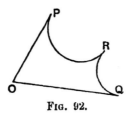

Fig. 92.

41*. A circle PBQ, centre A, passes through the centre B of a circle RST; if RP, SQ are common tangents, prove that PQ touches the circle RST.

42*. O is the centre of a fixed circle; two variable circles, centres P, Q, touch the fixed circle internally and each other externally; prove that the perimeter of △OPQ is constant.

43*. Two given circles touch internally at A; a variable line through A cuts the circles at P, Q; prove that the perpendicular bisector of PQ passes through a fixed point.

44*. OA, OB are two radii of a circle, such that ∠AOB = 60°; a circle touches OA, OB and the arc AB; prove that its radius = ⅓OA.

45*. C is the mid-point of AB; semicircles are drawn with AC, CB, AB as diameters and on the same side of AB; a circle is drawn to touch the three semicircles; prove that its radius = ⅓AC.

46*. A square ABCD is inscribed in a circle, and another square PQRS is inscribed in the minor segment AB; prove that AB = 5PQ.

CONVERSE PROPERTIES

Theorem 36

(i) If $\angle APB = \angle AQB$ and if P, Q are on the same side of AB, the four points A, B, P, Q lie on a circle.

(ii) If $\angle APB + \angle AQB = 180°$ and if P, Q are on opposite sides of AB, the four points A, B, P, Q lie on a circle.

(iii) If $\angle APB = 90°$, then P lies on the circle whose diameter is AB.

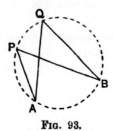
Fig. 93.
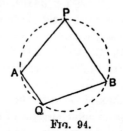
Fig. 94.

Converse of Theorem 41

If C and T are points on opposite sides of a line AB and such that $\angle BAT = \angle ACB$, then AT is a tangent to the circle which passes through A, C, B.

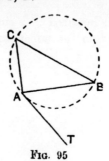
Fig. 95

CONVERSE PROPERTIES
EXERCISE XVII

1. **ABCD** is a parallelogram; if $\angle$ **ABC** $= 60°$, prove that the centre of the circle **ABD** lies on the circle **CBD**.
2. **BE, CF** are altitudes of $\triangle$**ABC**; prove that $\angle$ **AEF** $= \angle$ **ABC**.
3. The altitudes **AD, BE** of $\triangle$**ABC** intersect at **H**; prove that $\angle$ **DHC** $= \angle$ **DEC**.
4. **ABCD** is a parallelogram: any circle through **A, D** cuts **AB, DC** at **P, Q**; prove that **B, C, Q, P** are concyclic.
5. **ABC** is a $\triangle$ inscribed in a circle; **BE, CF** are altitudes of $\triangle$**ABC**; prove that **EF** is parallel to the tangent at **A**.
6. The circle **BCGF** lies inside the circle **ADHE**; **OABCD** and **OEFGH** are two lines cutting them; if **A, B, F, E** are concyclic, prove that **C, D, H, G** are concyclic.
7. **ABCD** is a parallelogram; **AC** cuts **BD** at **O**; prove that the circles **AOB, COD** touch each other.
8. A line **AD** is trisected at **B, C**; **BPC** is an equilateral triangle; prove that **AP** touches the circle **PBD**.
9. **AB** is a diameter, **AP** and **AQ** are two chords of a circle; **AP, AQ** cut the tangent at **B** in **X, Y**; prove that **P, X, Y, Q** are concyclic.
10. **ABC** is a $\triangle$ inscribed in a circle; any line parallel to **AC** cuts **BC** at **X**, and the tangent at **A** at **Y**; prove **B, X, A, Y** are concyclic.
11. In Fig. 96, **BQP** and **BAC** are equiangular isosceles triangles; prove that **QA** is parallel to **BC**.

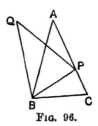

Fig. 96.

12. **ABCD** is a parallelogram; a circle is drawn touching **AD** at **A** and cutting **AB, AC** at **P, Q**; prove that **P, Q, C, B** are concyclic.

13. ABCD is a rectangle; the line through C perpendicular to AC cuts AB, AD produced at P, Q; prove that P, D, B, Q are concyclic.
14. In △ABC, ∠BAC = 90°; the perpendicular bisector of BC cuts CA, BA produced at P, Q; prove that BC touches the circle CPQ.
15. ABCDE is a regular pentagon; BD cuts CE at O; prove that BC touches the circle BOE.
16. OY is the bisector of ∠XOZ; P is any point; PX, PY, PZ are the perpendiculars to OX, OY, OZ; prove that XY = YZ.
17. AA^1, BB^1, CC^1 are equal arcs of a circle; AB cuts $A^1 B^1$ at P; AC cuts $A^1 C^1$ at Q; prove that A, A^1, P, Q are concyclic.
18. CA, CB are two fixed radii of a circle; P is a variable point on the circumference; PQ, PR are the perpendiculars from P to CA, CB; prove that QR is of constant length.
19. ABC is a △ inscribed in a circle; a line parallel to AC cuts BC at P, and the tangent at A at T; prove that ∠APC = ∠BTA.
20. O is a fixed point inside a given △ABC; X is a variable point on BC; the circles BXO, CXO cut AB, AC at Z, Y; prove that (1) O, Y, A, Z are concyclic, (2) the angles of △XYZ are of constant size.
21. Four circular coins of unequal sizes lie on a table so that each touches two, and only two, of the others; prove that the four points of contact are concyclic.
22. ABC, ABD are two equal circles; if AB = BC, prove that AC touches the circle ABD.
23. AB, CD are two intersecting chords of a circle; AP, CQ are the perpendiculars from A, C to CD, AB; prove that PQ is parallel to BD.
24. Prove that the quadrilateral formed by the external bisectors of any quadrilateral is cyclic.
25. AC, BD are two perpendicular chords of a circle; prove that the tangents at A, B, C, D form a cyclic quadrilateral.
26. AB, AC are two equal chords of a circle; AP, AQ are two chords cutting BC at X, Y; prove P, Q, X, Y are concyclic.
27. The diagonals of a cyclic quadrilateral ABCD intersect at

right angles at **O**; prove that the feet of the perpendiculars from **O** to **AB, BC, CD, DA** are concyclic.

28. **AOB, COD** are two perpendicular chords of a circle; **DE** is any other chord; **AF** is the perpendicular from **A** to **DE**; prove that **OF** is parallel to **BE**.
29. **ABC** is a $\triangle$ inscribed in a circle; **AD** is an altitude of $\triangle$**ABC**; **DP** is drawn parallel to **AB** and meets the tangent at **A** at **P**; prove $\angle$ **CPA** $= 90°$.
30. **BE, CF** are altitudes of $\triangle$**ABC**; **X** is the mid-point of **BC**; prove that **XE = XF**.
31. **BE, CF** are altitudes of $\triangle$**ABC**; **X** is the mid-point of **BC**; prove that $\angle$ **FXE** $= 180° - 2 \angle$ **BAC**.
32. Two circles **APRB, ASQB** intersect at **A, B**; **PAQ** and **RAS** are straight lines; **RP** and **QS** are produced to meet at **O**; prove that **O, P, B, Q** are concyclic.
33. **AOB, COD** are two perpendicular diameters of a circle; two chords **CP, CQ** cut **AB** at **H, K**; prove that **H, K, Q, P** are concyclic.
34. The side **CD** of the square **ABCD** is produced to **E**; **P** is any point on **CD**; the line from **P** perpendicular to **PB** cuts the bisector of $\angle$ **ADE** at **Q**; prove **BP = PQ**.
35*. **AB, CD** are parallel chords of a circle, centre **O**; **CA, DB** are produced to meet at **P**; the tangents at **A, D** meet at **T**; prove that **A, D, P, O, T** are concyclic.
36*. **X, Y** are the centres of the circles **ABP, ABQ**; **PAQ** is a straight line; **PX** and **QY** are produced to meet at **R**; prove that **X, Y, B, R** are concyclic.
37*. **BE, CF** are altitudes of $\triangle$**ABC**; **Z** is the mid-point of **AB**; prove that $\angle$ **ZEF** $= \angle$ **ABC** $\sim \angle$ **BAC**.
38*. **PQ** is a chord of a circle; the tangents at **P, Q** meet at **T**; **R** is any point such that **TR = TP**; **RP, RQ** cut the circle again at **E, F**; prove that **EF** is a diameter.
39*. **PQ, CD** are parallel chords of a circle; the tangent at **D** cuts **PQ** at **T**; **B** is the point of contact of the other tangent from **T**; prove that **BC** bisects **PQ**.
40*. **ABCD** is a parallelogram; **O** is a point inside **ABCD** such that $\angle$ **AOB** $+ \angle$ **COD** $= 180°$; prove that $\angle$ **OBC** $= \angle$ **ODC**.

MENSURATION

1. For a *circle* of radius r inches,
    (i) the length of the circumference $= 2\pi r$ in.
    (ii) the area of the circle $= \pi r^2$ sq. in.
    (iii) the length of an arc, which subtends $\theta°$ at the centre of the circle, $= \dfrac{\theta}{360} \times 2\pi r$ in.
    (iv) the area of a sector of a circle of angle $\theta° = \dfrac{\theta}{360} \times \pi r^2$ sq. in.
2. For a *sphere* of radius r inches,
    (i) the area of surface of sphere $= 4\pi r^2$ sq. in.
    (ii) the volume of the sphere $= \tfrac{4}{3}\pi r^3$ cub. in.
    (iii) the area of the surface intercepted between two parallel planes at distance d inches apart $= 2\pi r d$ sq. in.
3. For a *circular cylinder*, radius r inches, height h inches,
    (i) the area of the curved surface $= 2\pi r h$ sq. in.
    (ii) the volume of the cylinder $= \pi r^2 h$ cub. in.
4. For a *circular cone*, radius of base r inches, height h inches, length of slant edge l inches,
    (i) $l^2 = r^2 + h^2$.
    (ii) area of the curved surface $= \pi r l$ sq. in.
    (iii) volume of cone $= \tfrac{1}{3}\pi r^2 h$ cub. in.
5. (i) The volume of any cylinder = area of base × height.
    (ii) The volume of any pyramid = $\tfrac{1}{3}$ area of base × height.

$\pi = \tfrac{22}{7}$ approx. or $3\cdot1416$ approx.

MENSURATION

EXERCISE XVIII

1. Find (1) the circumference, (2) the area of a circle of radius (i) 4″, (ii) 100 yards.
2. The circumference of a circle is 5 inches; what is its radius correct to $\frac{1}{10}$ inch?
3. The area of a circle is 4 sq. cms.; what is its radius correct to $\frac{1}{10}$ cm.?
4. An arc of a circle of radius 3 inches subtends an angle of 40° at the centre; what is its length correct to $\frac{1}{10}$ inch?
5. The angle of a sector of a circle is 108°, and its radius is 2·5 cms.; what is its area?
6. A square **ABCD** is inscribed in a circle of radius 4 inches; what is the area of the minor segment cut off by **AB**.
7. **AB** is an arc of a circle, centre **O**; **AO** = 5 cms. and arc **AB** = 5 cms.; find ∠ **AOB**, correct to nearest minute.
8. A piece of flexible wire is in the form of an arc of a circle of radius 4·8 cms. and subtends an angle of 240° at the centre of the circle: it is bent into a complete circle: what is the radius?
9. A horse is tethered by a rope 5 yards long to a ring which can slide along a low straight rail 8 yards long; what is the area over which the horse can graze?
10. **OA**, **OB** are two radii of a circle; prove that the area of sector **AOB** equals $\frac{1}{2}$**OA** × arc **AB**.
11. What is the area contained between two concentric circles of radii 6 inches, 3 inches?

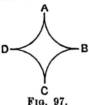

Fig. 97.

12. In Fig. 97, **AB**, **BC**, **CD**, **DA** are quadrants of equal circles of radii 5 cms., touching each other. Find the area of the figure.

13. Find (i) the volume, (ii) the *total* surface of a closed cylinder, height 8″, radius 5″.

14. 1 lb. of tobacco is packed in a cylindrical tin of diameter 4″ and height 8″; what would be the height of a tin of diameter 3″ which would hold ¼ lb. of tobacco, similarly packed?

15. How many cylindrical glasses 2″ in diameter can be filled to a depth of 3″ from a cylindrical jug of diameter 5″ and height 12″?

16. Find (i) the volume, (ii) the area of the curved surface of a circular cone, radius of base 5″, height 12″.

17. A sector of a circle of radius 5 cms. and angle 60° is bent to form the surface of a cone; find the radius of its base.

18. The curved surface of a circular cone, height 3″, radius of base 4″ is folded out flat. What is the angle of the sector so obtained?

19. Find (i) the volume, (ii) the *total* area of the surface of a pyramid, whose base is a square of side 6″ and whose height is 4″.

20. Find (i) the volume, (ii) the area of the surface of a sphere of diameter 5 cms.

21. Taking the radius of the earth as 4000 miles, find the area between latitudes 30° N and 30° S. What fraction is this area of the area of the total surface of the earth?

22. Two cylinders, diameters 8″ and 6″, are filled with water to depths 10″, 5″ respectively: they are connected at the bottom by a tube with a tap: when the tap is turned on, what is the resulting depth in each cylinder?

23. Three draughts, 1¼″ in diameter, are placed flat on a table and an elastic band is put round them. Find its stretched length.

24. What is the length of a belt which passes round two wheels of

MENSURATION

diameters 2", 4", so that the two straight portions cross at right angles? (see Fig. 98).

Fig. 98.

25. A circular metal disc, 9" in diameter, weighs 6 lb.; what is the weight of a disc of the same metal, 6" in diameter and of the same thickness?

26. Find the volume of the greatest circular cylinder that can be cut from a rectangular block whose edges are 4", 5", 6".

27. Fig. 99 (not drawn to scale) is a street plan, in which **EF** is a quadrant and the angles at **A, H, D, E, F** are 90°; **AE = AB = DF** = 100 yards; **HD** = 300 yards; **CH** = 150 yards. Find the two distances of **A** from **D** by the routes (i) **AEFD**, (ii) **ABCD**.

Find also the area in acres of the plot **ABCDFE**.

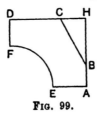

Fig. 99.

28. **AB, BC, CA** are three circular arcs, each of radius 6 cms. and touching each other at **A, B, C** (see Fig. 100)—
   (i) Calculate the area of the figure.
   (ii) Find its perimeter.

Fig. 100.

29. Draw a circle of radius 5 cms. and place in it a chord **AB** of length 4 cms.; find the area of the major segment **AB**, making any measurements you like.

30. A rectangular lawn 15 yards by 10 yards is surrounded by flower-beds: a man can, without stepping off the lawn, water the ground within a distance of 5 feet from the edge. What is the total area of the beds he can so water?

What would be the area within his reach, if the lawn was in the shape of (i) a scalene triangle, (ii) any convex polygon, of perimeter 50 yards?

31. **ABC** is a right-angled triangle; circles are drawn with **AB**, **BC**, **CA** as diameters; prove that the area of the largest is equal to the sum of the areas of the other two circles.

32. Fig. 101 represents four semicircles; **AC = DB** and **XOV** bisects **AB** at right angles. Prove that—
    (i) Curves **AXB**, **AVB** are of equal lengths;
    (ii) Area of figure = area of circle on **XV** as diameter.

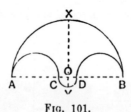

Fig. 101.

33. In Fig. 102, **BQA**, **APC**, **BSARC** are semicircles, prove that the sum of the areas of the lunes **BSAQ**, **CRAP** equals the area of △**ABC**.

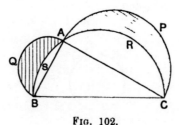

Fig. 102.

MENSURATION

34. In Fig. 103, **AB** = **BC** = **CA** = 2 cms., and the circular arcs touch the sides of △**ABC**; find the area of the figure.

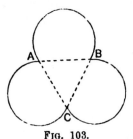

FIG. 103.

35*. A hoop, of radius 2', rests in a vertical position on a horizontal plane, with its rim in contact at **A** with a thin vertical peg, 1⁰ high. The hoop is rolled over the peg into the corresponding position on the other side: Fig. 104 shows the area thus swept out. Calculate this area.

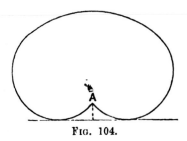

FIG. 104.

36*. A triangular piece of cardboard **ABC** is such that **BA** = 8", **AC** = 6", ∠**BAC** = 90°. It is placed on the floor with the edge **BC** against the wall and a pin is put through the midpoint of **BC**. The cardboard is now turned about **C** till **CA** is against the wall, then about **A** till **AB** is against the wall, then about **B** till **BC** is against the wall; the cardboard remains in contact with the floor throughout. Construct the curve which the pin scratches on the floor and find the area between this curve and the wall.

37*. The diagonals **AC**, **BD** of the quadrilateral **ABCD** cut at right angles at **O**; **AO** = 6", **OC** = **OD** = 2", **OB** = 4". The triangle **DOC** is cut away and the triangles **AOD**, **BOC** are folded through 90° about **OA**, **OB** so as to form two faces of a tetrahedron on △**OAB** as base.

Find (i) the volume of the tetrahedron;
(ii) the area of the remaining face;
(iii) the length of the perpendicular from **O** to the opposite face.

38*. **ABCD** is a rectangle; **AB** = 10″, **AD** = 6″; **AXB, BYC, CZD, DWA** are isosceles triangles, all the equal sides of which are 9″; they are folded so as to form a pyramid with **ABCD** as base and **X, Y, Z, W** at the vertex.

Find (i) the height of the pyramid;
(ii) the volume of the pyramid;
(iii) the *total* area of the surface of the pyramid.

If **AB** = p'', **AD** = q'', **AX** = r'', and if the height of the pyramid = h'', prove that $h^2 = r^2 - \frac{1}{4}p^2 - \frac{1}{4}q^2$.

LOCI

THEOREM 45

A, B are two fixed points; if a variable point **P** moves so that **PA = PB**, then the locus of (or path traced out by) **P** is the perpendicular bisector of **AB**.

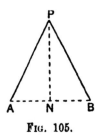

FIG. 105.

THEOREM 46

AOB, COD are two fixed intersecting lines; if a variable point **P** moves so that its perpendicular distances **PH, PK** from these lines are equal, then the locus of (or path traced out by) **P** is the pair of lines which bisect the angles between **AOB** and **COD**.

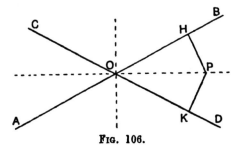

FIG. 106.

DEFINITION.—Given a point **P** and a line **AB**, if the perpendicular **PX** from **P** to **AB** is produced to **P¹** so that **PX = XP¹**, then **P¹** is called the *image* or *reflection* of **P** in **AB**.

LOCI

EXERCISE XIX

1. A variable point is at a given distance from a given line, what is its locus?
2. A variable point is at a given distance from a given point, what is its locus?
3. A variable circle touches a fixed line at a fixed point, what is the locus of its centre?
4. A variable circle passes through two fixed points, what is the locus of its centre?
5. A variable circle touches two fixed lines, what is the locus of its centre?
6. A variable circle of given radius passes through a fixed point, what is the locus of its centre?
7. A variable circle of given radius touches a fixed circle, what is the locus of its centre?
8. A variable circle touches two fixed concentric circles, what is the locus of its centre?
9. A variable circle of given radius touches a given line, what is the locus of its centre?
10. PQR is a variable triangle; $\angle$ QPR = 90°, PQ and PR pass through fixed points; what is the locus of P?
11. A, B are fixed points; APB is a triangle of given area; what is the locus of P?
12. Given the base and vertical angle of a triangle, find the locus of its vertex.
13. A variable chord of a fixed circle is of given length, what is the locus of its mid-point?
14. A is a fixed point on a fixed circle; AP is a variable chord; find the locus of the mid-point of AP.
15. P is a variable point on a given line; O is a fixed point outside the line; find the locus of the mid-point of OP.
16. A, B are fixed points; PAQB is a variable parallelogram of given area; find the locus of P.
17. ABC is a given triangle; BAPQ, CBQR are variable parallelo-

grams; if **P** moves on a fixed circle, centre **A**, find the locus of **R**.

18. A variable chord **PQ** of a given circle passes through a fixed point; find the locus of the mid-point of **PQ**.
19. The extremities of a line of given length move along two fixed perpendicular lines; find the locus of its mid point.
20. **A, B** are fixed points; **ABPQ** is a variable parallelogram; if **AP** is of given length, find the locus of **Q**.
21. **PQ, QR** are variable arcs of given lengths of a fixed circle, centre **O**; **PQ** meets **OR** at **S**; find the locus of **S**.
22. **O, A** are fixed points; **P** is a variable point on **OA**; **OPQ** is a triangle such that **OP + PQ** is constant and $\angle$ **OPQ** is constant; prove that the locus of **Q** is a straight line.
23. **PQR** is a variable triangle; the mid-points of **PQ** and **PR** are fixed and **QR** passes through a fixed point; find the locus of **P**.
24. **A, B** are fixed points; **P** moves along the perpendicular bisector of **AB**; **AP** is produced to **Q** so that **AP = PQ**; find the locus of **Q**.
25. **A, B** are fixed points; **P** is a variable point such that $AP^2 + PB^2$ is constant; find the locus of **P**.
26. **A, B** are fixed points; **P** is a variable point such that $PA^2 - PB^2$ is constant; prove that the locus of **P** is a straight line perpendicular to **AB**.
27. **AB, AC** are two fixed lines; **P** is a variable point inside $\angle$ **BAC** such that the sum of its distances from **AB** and **AC** is constant; prove that the locus of **P** is a straight line.
28. **A, B, C, D** are fixed points; **P** is a variable point such that the sum of the areas of the triangles **PAB, PCD** is constant; prove that the locus of **P** is a straight line.
29. If **P¹** is the image of **P** in the line **AB**, prove that **AP = AP¹**.
30. A variable line **OQ** passes through a fixed point **O**; **A** is another fixed point; find the locus of the image of **A** in **OQ**.
31. **A, B** are two points on the same side of a line **CD**; **A¹** is the image of **A** in **CD**; **A¹B** cut **CD** at **O**; prove that—
    (i) **AO** and **OB** make equal angles with **CD**;
    (ii) if **P** is any other point on **CD**, **AP + PB > AO + OB**.

32. **AH, BK** are the perpendiculars from **A, B** to **XY**. AH = 5″, BK = 7″, HK = 16″; what is the least value of **AP + PB** ?

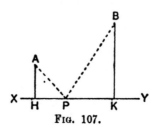

Fig. 107.

33. **A, B** are fixed points on opposite sides of a fixed line **CD**; find the point **P** on **CD** for which **PA∼PB** has its greatest value.
34. How many images are formed when a candle is placed between two plane mirrors inclined to each other at an angle of (i) 90°; (ii) 60°?
35. If a billiard ball at **A** moves so as to hit a perfectly elastic cushion **XY** at **P**, it will continue in the line **A¹PB** where **A¹** is the image of **A** in **XY**; or, in other words, the two portions of its path **AP** and **PB** make equal angles with **XY**. **ABCD** is a rectangular billiard table with perfectly elastic cushions: a ball is at any point **P**; it is struck in a direction parallel to **AC**; prove that after hitting all four cushions it will again pass through **P**.

THE TRIANGLE—CONCURRENCY PROPERTIES

Theorem 47

If **ABC** is a triangle, the perpendicular bisectors of **BC. CA, AB** meet at a point **O** (say).

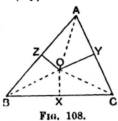

Fig. 108.

O is the centre of the circumcircle of the triangle **ABC**, and is called the *circumcentre*.

Theorem 48

If **ABC** is a triangle, the internal bisectors of the angles **ABC, BCA, CAB** meet at a point **I** (say).

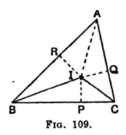

Fig. 109.

I is the centre of the circle inscribed in the triangle **ABC** (*i.e.* the in-circle of △**ABC**), and is called the *in-centre*. The external bisectors of the angles **ABC, ACB** meet at a point I_1, which is the centre of the circle which touches **AB** produced, **AC** produced, **BC**; this circle is said to be *escribed* to **BC**, and I_1 is called an *ex-centre*.

7

Theorem 49

If **ABC** is a triangle, the altitudes **AD, BE, CF** meet at a point **H** (say).

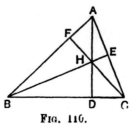

Fig. 110.

H is called the *orthocentre* of the triangle **ABC**. The triangle **DEF** is called the *pedal triangle* of △**ABC**.

Theorem 50

If **ABC** is a triangle, the medians **AD, BE, CF** meet at a point **G** (say), and **DG** = $\frac{1}{3}$**DA**.

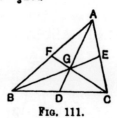

Fig. 111.

G is called the *centroid* of the triangle **ABC**.

THE TRIANGLE—CONCURRENCY PROPERTIES

EXERCISE XX

THE CIRCUMCIRCLE

1. If **O** is the circumcentre of $\triangle$**ABC** and if **D** is the mid-point of **BC**, prove $\angle$ **BOD** = $\angle$ **BAC**.
2. The diagonals of the quadrilateral **ABCD** intersect at **O**; **P, Q, R, S** are the circumcentres of $\triangle$s **AOB, BOC, COD, DOA**; prove **PQ** = **RS**.
3. In $\triangle$**ABC**, $\angle$ **BAC** = 90°; **P** is the centre of the square described on **BC**; prove that **AP** bisects $\angle$ **BAC**.
4. In $\triangle$**ABC**, $\angle$ **BAC** = 90°; prove that the perpendicular bisectors of **AB** and **AC** meet on **BC**.
5. **ABC** is a scalene triangle; prove that the perpendicular bisector of **BC** and the bisector of $\angle$ **BAC** meet *outside* the triangle **ABC**.
6. **ABCD** is a parallelogram; **E, F** are the circumcentres of $\triangle$s **ABD, BCD**; prove that **EBFD** is a rhombus.
7. The extremities of a variable line **PQ** of given length lie on two fixed lines **OA, OB**; prove that the locus of the circumcentre of $\triangle$**OPQ** is a circle, centre **O**.
8. If the area of the triangle **ABC** is $\triangle$, the radius of the circumcircle is $\dfrac{abc}{4\triangle}$; prove this for the case where $\angle$ **BAC** = 90°.
9. **ABCD** is a quadrilateral such that **AB** = **CD**; find a point **O** such that $\triangle$**OAB** $\equiv$ $\triangle$**OCD**.
10. **AD, BE** are altitudes of $\triangle$**ABC**; prove that the perpendicular bisectors of **AD, BE, DE** are concurrent.
11. In $\triangle$**ABC**, **AB** = **AC**; **P** is any point on **BC**; **E, F** are the circumcentres of $\triangle$s **ABP, ACP**; prove that **AE** is parallel to **PF**.

THE IN-CIRCLE AND EX-CIRCLES

12. In Fig. 112, if $BC = a$, $CA = b$, $AB = c$, and $s = \tfrac{1}{2}(a+b+c)$ prove that
    (i) $AY = s - a$.
    (ii) $AQ = s$.
    (iii) $BP = XC$.
    (iv) $YQ = ZR$.
    (v) $XP = b \sim c$.
    (vi) $IX = \dfrac{\triangle}{s}$ where $\triangle$ = area of triangle ABC.
    (vii) $I_1 P = \dfrac{\triangle}{s-a}$.
    (viii) B, I, C, I_1 are concyclic.
    (ix) $AZ + BX + CY = s$.
    (x) if $\angle BIC = 100°$, calculate $\angle BAC$.

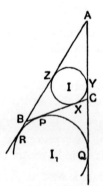

Fig. 112.

13. AB is a chord of a circle; the tangents at A, B meet at T; prove that the in-centre of $\triangle TAB$ lies on the circle.
14. I is the in-centre and O the circumcentre of $\triangle ABC$; prove that $\angle IAO = \tfrac{1}{2}(ABC \sim \angle ACB)$.
15. I is the in-centre of $\triangle ABC$; prove that $\angle AIC = 90° + \tfrac{1}{2}\angle ABC$.
16. I is the in-centre and AD is an altitude of $\triangle ABC$; prove that $\angle IAD = \tfrac{1}{2}(\angle ABC \sim \angle ACB)$.
17. In Fig. 112, prove that $AB - AC = BX - XC$.

THE TRIANGLE—CONCURRENCY PROPERTIES

18. The in-circle of $\triangle ABC$ touches BC at X, prove that the in-circles of $\triangle$s ABX, ACX touch each other.
19. ABCD is a quadrilateral circumscribing a circle; prove that the in-circles of $\triangle ABC$, CDA touch each other.
20. Two concentric circles are such that a triangle can be inscribed in one and circumscribed to the other; prove that the triangle is equilateral.
21. In $\triangle ABC$, $\angle BAC = 90°$; prove that the diameter of the in-circle of $\triangle ABC$ equals $AB + AC - BC$.
22. The extremities P, Q of a variable line lie on two fixed lines AB, CD; the bisectors of $\angle$s APQ, CQP meet at R; find the locus of R.
23. I is the in-centre of $\triangle ABC$; I_1 is the centre of the circle escribed to BC; I, I_1 cuts the circumcircle of $\triangle ABC$ at P; prove that I, I_1, B, C lie on a circle, centre P.
24. I is the in-centre of $\triangle ABC$; if the circumcircle of $\triangle BIC$ cuts AB at Q, prove $AQ = AC$.
25. I is the in-centre of $\triangle ABC$; AP, AQ are the perpendiculars from A to BI, CI; prove that PQ is parallel to BC.
26*. The in-circle of $\triangle ABC$ touches BC, CA at X, Y; I is the in-centre; XY meets AI at P; prove $\angle BPI = 90°$.

THE ORTHOCENTRE

27. If AD, BE, CF are the altitudes of $\triangle ABC$ and if H is its orthocentre (see Fig. 110), prove that
    (i) $\angle BHF = BAC$.
    (ii) $\angle BHC + \angle BAC = 180°$.
    (iii) $\triangle$s AEF, ABC are equiangular.
    (iv) $\triangle$s BDF, EDC are equiangular.
    (v) AD bisects $\angle FDE$.
    (vi) $\angle EDF = 180° - 2\angle BAC$.
    (vii) H is in-centre of $\triangle DEF$.
28. Where is the orthocentre of a right-angled triangle?
29. Q is a point inside the parallelogram ABCD such that $\angle QBC = 90° = \angle QDC$; prove that AQ is perpendicular to BD.
30. If D is the orthocentre of $\triangle ABC$, prove that A is the orthocentre of $\triangle BCD$.

31. If H is the orthocentre of $\triangle ABC$, prove that the circumcircles of $\triangle$s AHB, AHC are equal.
32. I is the in-centre and I_1, I_2, I_3 are the ex-centres of $\triangle ABC$, prove that I_1 is the orthocentre of $\triangle I I_2 I_3$.
33. In $\triangle ABC$, $AB = AC$, $\angle BAC = 45°$; H is the orthocentre a: CHF is an altitude; prove that $BF = FH$.
34. O is the circumcentre and H the orthocentre of $\triangle ABC$; prove that $\angle HBA = \angle OBC$.
35. P, Q, R are the mid-points of BC, CA, AB; prove that the orthocentre of $\triangle PQR$ is the circumcentre of $\triangle ABC$.
36. H is the orthocentre of $\triangle ABC$; AH meets BC at D and the circumcircle of $\triangle ABC$ at P; prove that $HD = DP$.
37. O is the circumcentre, I is the in-centre, H is the orthocentre of $\triangle ABC$; prove that AI bisects $\angle OAH$.
38. BE, CF are altitudes of $\triangle ABC$; O is its circumcentre; prove that OA is perpendicular to EF.
39. H is the orthocentre and O the circumcentre of $\triangle ABC$; AK is a diameter of the circumcircle; prove that (i) BHCK is a parallelogram, (ii) CH equals twice the distance of O from AB.
40*. H is the orthocentre and O the circumcentre of $\triangle ABC$; if $AO = AH$, prove $\angle BAC = 60°$.
41. H is the orthocentre of $\triangle ABC$; BH meets the circumcircle at K; prove $AH = AK$.
42*. The altitudes BE, CF of $\triangle ABC$ meet at H; P, X are the mid-points of AH, BC; prove that PX is perpendicular to EF.
43. Given the base and vertical angle of a triangle, find the locus of its orthocentre.
44. [*Nine Point Circle.*] AD, BE, CF are altitudes of $\triangle ABC$; H is its orthocentre; X, Y, Z, P, Q, R are the mid-points of BC, CA, AB, HA, HB, HC; prove that
   (i) PZ is parallel to BE and ZX is parallel to AC.
   (ii) $\angle PZX = 90°$ and $\angle PYX = 90°$.
   (iii) P, Z, X, D, Y lie on a circle.
   (iv) The circle through X, Y, Z passes through P, Q, R, D, E, F.

THE CENTROID

45. **X, Y, Z** are the mid-points of **BC, CA, AB**; prove that the triangles **ABC, XYZ** have the same centroid.
46. **ABCD** is a parallelogram; **P** is the mid-point of **AB**; **CP** cuts **BD** at **Q**; prove that **AQ** bisects **BC**.
47. If the medians **AX, BY** of $\triangle$**ABC** meet at **G**, prove that $\triangle$s **BGX, CGY** are equal in area.
48. If **G** is the centroid of $\triangle$**ABC** and if **AG** = **BC**, prove that $\angle$ **BGC** = $90°$.
49. If two medians of a triangle are equal, prove that the triangle is isosceles.
50. **X, Y, Z** are the mid-points of **BC, CA, AB**; **AD** is an altitude of $\triangle$**ABC**; prove that $\angle$ **ZXY** = $\angle$ **ZDY** = $\angle$ **BAC**.
51. **AX, BY, CZ** are the medians of $\triangle$**ABC**; prove that **BY** + **CZ** > **AX**.
52. If the centroid and circumcentre of a triangle coincide, prove that the triangle is equilateral.
53. **ABCD** is a parallelogram; **H, K** are the mid-points of **AB, AD**; prove that **CH** and **CK** trisect **BD**.
54*. In a tetrahedron **ABCD**, the plane angles at each of three corners add up to $180°$; prove, by drawing the net of the tetrahedron, that its opposite edges are equal.

RIDERS ON BOOK IV

PROPORTION

THEOREM 51

If the heights **AP, XQ** of the triangles **ABC, XYZ** are equal,
$$\frac{\triangle ABC}{\triangle XYZ} = \frac{BC}{YZ}.$$

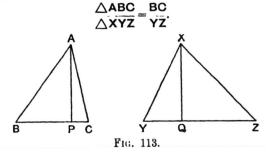

FIG. 113.

THEOREM 52

(1) If a straight line, drawn parallel to the base **BC** of the triangle **ABC**, cuts the sides **AB, AC** (produced if necessary) at **H, K**, then $\frac{AH}{HB} = \frac{AK}{KC}$ and $\frac{AH}{AB} = \frac{AK}{AC}$.

(2) If **H, K** are points on the sides **AB, AC** (or the sides produced) of the triangle **ABC** such that $\frac{AH}{HB} = \frac{AK}{KC}$, then **HK** is parallel to **BC**.

FIG. 114(1).

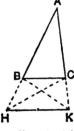

FIG. 114(2).

FIG. 114(3).

PROPORTION

EXERCISE XXI

1. What is the value of the following ratios:
   (i) 3 ins. : 2 ft.; (ii) $4d. : 2s.$; (iii) 20 min. : $1\frac{1}{2}$ hr.; (iv) 3 sq. ft. : 2 sq. yd.; (v) 3 right angles : $120°$; (vi) 3 m. : 25 cms.?

2. Find x in the following:
   (i) $3 : x = 4 : 10$, (ii) x feet : 5 yards $= 2 . 3$; (iii) $6 : x = x : 24$; (iv) 2 hours : 50 minutes $= 3$ shillings : x shillings.

3. If $\dfrac{a}{b} = \dfrac{c}{d}$, prove that

   (i) $\dfrac{b}{a} = \dfrac{d}{c}$;   (ii) $ad = bc$,   (iii) $\dfrac{a+b}{b} = \dfrac{c+d}{d}$;

   (iv) $\dfrac{a+b}{a-b} = \dfrac{c+d}{c-d}$;   (v) $\dfrac{b+d}{a+c} = \dfrac{b}{a}$.

4. If $\dfrac{a}{b} = \dfrac{c}{d} = \dfrac{e}{f}$, fill up the blank spaces in the following:

   (i) $\dfrac{a}{a+b} = \dfrac{c}{}$;   (ii) $\dfrac{a-b}{a} = \dfrac{c-d}{}$;   (iii) $\dfrac{a+c}{b+d} = \dfrac{}{d+f}$;

   (iv) $\dfrac{a}{b} = \dfrac{}{b+d+f}$; (v) $\dfrac{a-3c}{b-3d} = \dfrac{2a+7c-23e}{}$; (vi) $\dfrac{ac}{bd} = \dfrac{a^2+e^2}{}$.

5. Solve the equations (i) $\dfrac{x+\frac{1}{2}}{x-\frac{1}{2}} = \dfrac{7}{3}$; (ii) $\dfrac{5x^2-3x+2}{5x^2+3x-2} = \dfrac{5x-1}{5x+1}$.

6. Are the following in proportion (i) $3\frac{1}{3}$, 5, 8, 12; (ii) 8 inches, 6 degrees, 12 degrees, 9 inches?

7. Find the fourth proportional to (i) 2, 3, 4; (ii) ab, bc, cd.

8. Find the third proportional to (i) $\frac{1}{2}$, $\frac{1}{5}$; (ii) x, xy.

9. Find a mean proportional between (i) 4, 25; (ii) a^2b, bc^2.

10. A line **AB**, 8″ long, is divided internally at **P** in the ratio $2 : 3$; find **AP**.

11. A line **AB**, 8″ long, is divided externally at **Q** in the ratio $7 : 3$; find **BQ**.

12. **AB** is divided internally at **C** in the ratio $5 : 6$. Is **C** nearer to **A** or **B**?

13. **AB** is divided externally at **D** in the ratio $9 : 7$. Is **D** nearer to **A** or **B**?

PROPORTION

14. **AB** is divided externally at **D** in the ratio $3:5$. Is **D** nearer to **A** or **B**?
15. A line **AB**, 6" long, is divided internally at **P** in the ratio $2:1$, and externally at **Q** in the ratio $5:2$; find the ratios in which **PQ** is divided by **A** and **B**.
16. **ABCDE** is a straight line such that $AB:BC:CD:DE = 1:3:2:5$. Find the ratios (i) $\dfrac{AB}{AE}$; (ii) $\dfrac{AC}{CE}$, (iii) $\dfrac{EB}{AD}$.
  Find the ratios in which **BE** is divided by **A** and **D**.
  If $BE = 4''$, find **AC**.
17. A line **AB**, 8" long, is divided internally at **C** and externally at **D** in the ratio $7:3$; **O** is the mid-point of **AB**, prove that $OC \cdot OD = OB^2$.
18. A line **AB**, 6" long, is divided internally at **C** and externally at **D** in the ratio $4:1$, **O** is the mid-point of **CD**, prove that $AO = 16 BO$, and find the length of **CD**.
19. A line of length x'' is divided internally in the ratio $a:b$; find the lengths of the parts.
20. A line of length y'' is divided externally in the ratio $a:b$; find the lengths of the parts.
21. A line **AB** is bisected at **O** and divided at **P** in the ratio $x:y$; find the ratio $\dfrac{OP}{AB}$.
22. **AB** is divided internally at **C** and externally at **D** in the ratio $x:y$; find (i) $\dfrac{CD}{AB}$, (ii) the ratio in which **B** divides **CD**.
23. **ABCDEF** is a straight line such that $AB:BC:CD:DE:EF = p:q:r:s:t$; find (i) $\dfrac{AB}{AF}$, (ii) $\dfrac{BE}{CF}$, (iii) the ratios in which **A** and **E** divide **CF**. If $BD = x''$, find **AE**.
24. **ABCD**, **AXYZ** are two straight lines such that $AB:BC:CD = AX:XY:YZ$. Fill up the blank spaces in the following:
  (i) $\dfrac{AB}{AX} = \dfrac{AC}{---}$; (ii) $\dfrac{BC}{AD} = \dfrac{}{AZ}$; (iii) $\dfrac{XZ}{AY} = \dfrac{}{AC}$.
25. **ABC** is a straight line; if $AC = \lambda \cdot AB$, find $\dfrac{AB}{BC}$ in terms of λ.
26. The sides of a triangle are in the ratio $x:y:z$ and its perimeter is p inches; find the sides.

27. **AB** is parallel to **CD**; **OB** = 2″, **OD** = 2½″, **BC** = 5″; find **AD**.

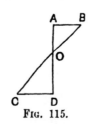

Fig. 115.

28. **AB**, **CD**, **EF** are parallel lines; **AC** = 2″, **CE** = 3″, **BF** = 4″; find **BD**.

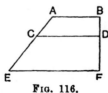

Fig. 116.

29*. $\dfrac{AG}{GB} = \dfrac{\lambda}{\mu}$; **AP**, **BQ**, **GN** are perpendicular to **OX**; **OP** = a, **OQ** = b; find **ON**.

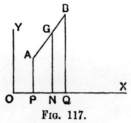

Fig. 117.

30*. The medians of $\triangle$**ABC** intersect at **G**; **AP**, **BQ**, **CR**, **GN** are the perpendiculars from **A**, **B**, **C**, **G** to a line **OX**; if **OP** = a, **OQ** = b, **OR** = c; prove **ON** = $\frac{1}{3}(a+b+c)$.

31. **ABC** is a $\triangle$; **P**, **Q** are points on **AB**, **AC** such that **AP** = ⅓**AB** and **CQ** = ⅓**CA**; prove that a line through **C** parallel to **PQ** bisects **AB**.

32. Three parallel lines **AX**, **BY**, **CZ** cut two lines **ABC**, **XYZ**; prove that $\dfrac{AB}{BC} = \dfrac{XY}{YZ}$

PROPORTION

33. The diagonals of the quad. ABCD intersect at O; if AB is parallel to DC, prove $\dfrac{AO}{AC} = \dfrac{BO}{BD}$.

34. A line parallel to BC cuts AB, AC at H, K; prove that AH . AC = AK . AB.

35. O is any point inside the △ABC; a line XY is drawn parallel to AB cuts OA, OB at X, Y; YZ is drawn parallel to BC to cut OC at Z; prove XZ is parallel to AC.

36. ABCD is a quadrilateral; P is any point on AB; lines PX, PY are drawn parallel to AC, AD to cut BC, BD at X, Y; prove XY is parallel to CD.

37. D is the foot of the perpendicular from A to the bisector of ∠ABC; a line from D parallel to BC cuts AC at X; prove AX = XC.

38. In Fig. 118, prove $\dfrac{\triangle ABC}{\triangle ABD} = \dfrac{CO}{OD}$.

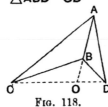

FIG. 118.

39. I is the in-centre of △ABC; prove that △IBC : △ICA : △IAB = BC : CA : AB.

40. In Fig. 118, prove $\dfrac{\triangle ACD}{\triangle BCD} = \dfrac{AO}{BO}$.

41*. In Fig. 119, AH = HB, AK = 2KC; find the ratio of the areas of the small triangles in the figure; hence find the ratio $\dfrac{CO}{OH}$.

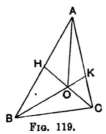

FIG. 119.

42*. ABC is a $\triangle$; H, K are points on AB, AC such that $HB = \frac{1}{4}AB$ and $KC = \frac{1}{3}AC$; BK cuts CH at O; prove $BO = OK$ and $CO = 2OH$. [Use method of ex. 41.]

43*. ABC is a $\triangle$; Y, Z are points on AC, AB such that $CY = \frac{1}{3}CA$ and $AZ = \frac{1}{2}ZB$; BY cuts CZ at O; prove $OY = \frac{1}{7}BY$ and $OZ = \frac{4}{7}CZ$. [Use method of ex. 41.]

44. Two circles APQ, AXY touch at A; APX, AQY are straight lines; prove $\dfrac{AP}{PX} = \dfrac{AQ}{QY}$.

45. ABCD is a parallelogram; any line through C cuts AB produced, AD produced at P, Q; prove $\dfrac{AB}{BP} = \dfrac{QD}{DA}$.

46*. ABCD is a parallelogram; a line through C cuts AB, AD, BD (produced if necessary) at P, Q, O; prove $OP \cdot OQ = OC^2$.

47. ABC is a $\triangle$; three parallel lines AP, BQ, CR meet BC, CA, AB (produced if necessary) at P, Q, R; prove that $\dfrac{BP}{PC} \times \dfrac{CQ}{QA} \times \dfrac{AR}{RB} = 1$.

48*. O is any point inside $\triangle ABC$; D, E, F are points on BC, CA, AB such that $AD = BE = CF$; lines are drawn from O parallel to AD, BE, CF to meet BC, CA, AB at P, Q, R; prove $OP + OQ + OR = AD$.

49*. ABC is a triangle; a line cuts BC produced, CA, AB at P, Q, R; CX is drawn parallel to PQ, meeting AB at X; prove (i) $\dfrac{BP}{PC} = \dfrac{BR}{RX}$; (ii) $\dfrac{BP}{PC} \times \dfrac{CQ}{QA} \times \dfrac{AR}{RB} = 1$.

[This is known as *Menelaus'* Theorem.]

SIMILAR TRIANGLES

THEOREM 53

If the triangles ABC, XYZ are equiangular ($\angle$ ABC = $\angle$ XYZ and $\angle$ ACB = $\angle$ XZY),

then $\dfrac{AB}{XY} = \dfrac{BC}{YZ} = \dfrac{CA}{ZX}$.

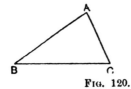

FIG. 120.

THEOREM 54

If the triangles ABC, XYZ are such that $\dfrac{AB}{XY} = \dfrac{BC}{YZ} = \dfrac{CA}{ZX}$, then the triangles are equiangular, $\angle$ ABC = $\angle$ XYZ, $\angle$ ACB = $\angle$ XZY, $\angle$ BAC = $\angle$ YXZ.

THEOREM 55

If, in the triangles ABC, XYZ, $\angle$ BAC = $\angle$ YXZ and $\dfrac{AB}{XY} = \dfrac{AC}{XZ}$, then the triangles are equiangular, $\angle$ ABC = $\angle$ XYZ and $\angle$ ACB = $\angle$ XZY.

SIMILAR TRIANGLES

EXERCISE XXII

1. A pole 10′ high casts a shadow $3\frac{1}{4}$′ long; at the same time a church spire casts a shadow 42′ long. What is its height?
2. In a photograph of a chest of drawers, the height measures 6″ and the breadth 3·2″; if its height is $7\frac{1}{2}$ feet, what is its breadth?
3. Show that the triangle whose sides are 5·1″, 6·8″, 8·5″ is right-angled.
4. A halfpenny (diameter 1″) at the distance of 3 yards appears nearly the same size as the sun or moon at its mean distance. Taking the distance of the sun as 93 million miles, find its diameter. Taking the diameter of the moon as 2160 miles, find its mean distance.
5. How far in front of a pinhole camera must a man 6′ high stand in order that a full-length photograph may be taken on a film $2\frac{1}{4}$″ high, $2\frac{1}{4}$″ from the pinhole?
6. The slope of a railway is marked as 1 in 60. What height (in feet) does it climb in $\frac{3}{4}$ mile?
7. A light is 9′ above the floor; a ruler, 8″ long, is held horizontally 4′ above the floor; find the length of its shadow.
8. Two triangles are equiangular; the sides of one are 5″, 8″, 9″; the shortest side of the other is 4 cms.; find its other sides.
9. The bases of two equiangular triangles are 4″, 6″; the height of the first is 5″; find the area of the second.
10. In △ABC, AB = 8″, BC = 6″, CA = 5″; a line XY parallel to BC cuts AB, AC at X, Y; AX = 2″; find XY, CY.
11. In quadrilateral ABCD, AB is parallel to DC and AB = 8″, AD = 3″, DC = 5″; AD, BC are produced to meet at P; find PD.
12. A line parallel to BC meets AB, AC at X, Y; BC = 8″, XY = 5″; the lines BC, XY are 2″ apart. Find the area of △ AXY.

SIMILAR TRIANGLES

13. In Fig. 121,
    (i) if $AO = 3''$, $OB = 2''$, $AB = 4''$, $DC = 1\frac{1}{2}''$, find CO, DO.
    (ii) if $AO = 5''$, $BO = 4''$, $AC = 7''$, find BD.
    (iii) if $PA = 9''$, $PB = 8''$, $AB = 4''$, $PC = 3''$, find PD, CD.
    (iv) if $PA = 9''$, $PB = 8''$, $AC = 6''$, $PC = 4''$, find BD, D.

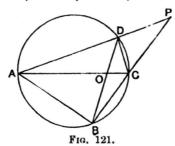

Fig. 121.

14. Show that the line joining (1, 1) to (4, 2) is parallel to and half of the line joining (0, 0) to (6, 2).
15. Three lines APB, AQC, ARD are cut by two parallel lines PQR, BCD; $AR = 3''$, $RD = 2''$, $BC = 4''$; find PQ.
16. In Fig. 122, AB is parallel to OD; $AB = 6'$, $BO = 20'$, $BE = 5'$, $DQ = 9'$; find OD, BP.

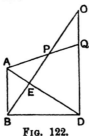

Fig. 122.

17. The diameter of the base of a cone is $9''$ and its height is $15''$; find the diameter of a section parallel to the base and $3''$ from it.
18. AXB is a straight line; AC, XY, BD are the perpendiculars from A, X, B to a line CD; $AC = 10$, $BD = 16$, $AX = 12$, $XB = 6$; find XY.
19. A, B are points on the same side of a line OX and at distances $1''$, $5''$ from it; Q and R divide AB internally and externally in the ratio 5 : 3; find the distances of Q and R from OX.

20. A rectangular table, 5' wide, 8' long, 3' high, stands on a level floor under a hanging lamp; the shadow on the floor of the shorter side is 8' long; find the length of shadow of the longer side and the height of the lamp above the table.

21. A sphere of 5″ radius is placed inside a conical funnel whose slant side is 12″ and whose greatest diameter is 14″; find the distance of the vertex from the centre of the sphere.

22. The length of each arm of a pair of nutcrackers is 6″; find the distance between the ends of the arms when a nut 1″ in diameter is placed with its nearer end 1″ from the apex.

23. In Fig. 123, **PQBR** is a rectangle.
    (i) If **AB** = 7, **PQ** = 1, **PR** = 2, find **BC**.
    (ii) If **AB** = 7, **BC** = 5, **PR** = x, **PQ** = y, find an equation between x, y.

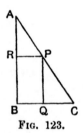

Fig. 123.

24. In △**ABC**, ∠**ABC** = 90°, **AB** = 5″, **BC** = 2″; the perpendicular bisector of **AC** cuts **AB** at **Q**; find **AQ**.

25. The diameter of the base of a cone is 8″; the diameter of a parallel section, 3″ from the base, is 6″; find the height of the cone.

26. In Fig. 124, **AB**, **PN**, **DC** are parallel; **AB** = 4″, **BC** = 5″, **CD** = 3″; calculate **PN**.

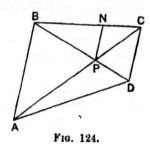

Fig. 124.

SIMILAR TRIANGLES

27. **ABCD** is a quadrilateral such that $\angle$ **ABC** $= 90° = \angle$ **ACD**, **AC** $= 5''$, **BC** $= 3''$, **CD** $= 10''$; calculate the distances of **D** from **BC, BA**.

28. **PQ** is a chord of a circle of length 5 cms.; the tangents at **P, Q** meet at **T**; **PR** is a chord parallel to **TQ**; if **PT** $= 8$ cms., find **PR**.

29. (i) A man, standing in a room opposite to and 6' from a window 27'' wide, sees a wall parallel to the plane of the window. With one eye shut, he can see 18'' less length of wall than with both eyes open; supposing his eyes are 2'' apart, find the distance of the wall from the window and the total length of wall visible.

    (ii) If the window is covered by a shutter containing a vertical slit $\frac{1}{2}''$ wide, show that there is a part of the wall out of view which lies between two parts in view and find its length.

    (iii) A man in bed at night sees a star pass slowly across a vertical slit in the blind; shortly afterwards, this occurs again. Is it possible that he sees the same star twice? Explain your answer by a figure.

30. A rectangular sheet of paper **ABCD** is folded so that **D** falls on **B**; the crease cuts **AB** at **Q**; **AB** $= 11''$, **AD** $= 7''$; find **AQ**.

31. Fig. 125 represents an object **HK** and its image **PQ** in a concave mirror, centre **O**, focus **F**.

    CH $= u$, **CP** $= v$, **CF** $= $ **FO** $= f$, **HK** $= x$, **PQ** $= y$;

    prove that (i) $\dfrac{1}{f} = \dfrac{1}{u} - \dfrac{1}{v}$; (ii) $y = \dfrac{vx}{u}$.

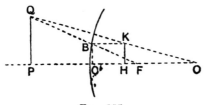

Fig. 125.

32. In Fig. 126, with the same notation as in ex. 31, prove that $\dfrac{1}{f} = \dfrac{1}{u} + \dfrac{1}{v}$, and find y in terms of x, u, f.

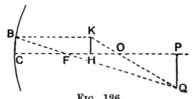

Fig. 126.

33. Fig. 127 represents an object HK and its image PQ in a thin concave lens, centre O, focus F.
OH $= u$, OP $= v$, OF $= f$, HK $= x$, PQ $= y$;
prove that (i) $\dfrac{1}{f} = \dfrac{1}{v} - \dfrac{1}{u}$; (ii) $y = \dfrac{vx}{u}$.

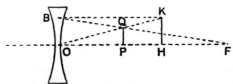

Fig. 127.

34. Fig. 128 represents an object HK and its image PQ in a thin convex lens, centre O, focus F.
OH $= u$, OP $= v$, OF $= f$, HK $= x$, PQ $= y$;
prove that $\dfrac{1}{f} = \dfrac{1}{u} + \dfrac{1}{v}$, and find y in terms of x, u, f.

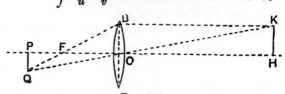

Fig. 128.

35. **OACB** is a quadrilateral on level ground; $\angle$ AOB $= 90° = \angle$ OBC, $\angle$ OAC $= 135°$, OB $= 9'$, OA $= 12'$; it is covered by a plane roof resting on pillars; the pillars at **A, B** are 10' high, the pillar at **O** is 8' high; find the height of the pillar at **C**.

SIMILAR TRIANGLES

36. **AB, DC** are the parallel sides of a trapezium **ABCD**; the diagonals cut at **O**; prove $\dfrac{AO}{OC} = \dfrac{AB}{CD}$.

37. **BE, CF** are altitudes of $\triangle ABC$; prove $\dfrac{BE}{CF} = \dfrac{AB}{AC}$.

38. **AOB, COD** are two intersecting chords of a circle; fill up the blank spaces in (i) $\dfrac{OA}{AC} = \dfrac{\ \ }{BD}$; (ii) $\dfrac{OA}{OC} = \dfrac{\ \ }{\ \ }$.

39. Two straight lines **OAB, OCD** cut a circle at **A, B, C, D**; fill up the blank spaces in (i) $\dfrac{AC}{BD} = \dfrac{OA}{\ \ }$; (ii) $\dfrac{OA}{OC} = \dfrac{\ \ }{\ \ }$.

40. **ABC** is a $\triangle$ inscribed in a circle; the bisector of $\angle$ **BAC** cuts **BC** at **Q** and the circle at **P**; prove $\dfrac{AC}{AP} = \dfrac{AQ}{AB}$ and complete the equation $\dfrac{BQ}{AB} = \dfrac{PC}{\ \ }$.

41. In $\triangle ABC$, $\angle BAC = 90°$; **AD** is an altitude; prove that $\dfrac{DC}{AC} = \dfrac{AC}{BC}$ and complete the equation $\dfrac{CD}{DA} = \dfrac{\ \ }{DB}$.

42. The medians **BY, CZ** of $\triangle ABC$ intersect at **G**; prove that $GY = \tfrac{1}{3} BY$.

43. **BE, CF** are altitudes of $\triangle ABC$; prove that $\dfrac{EF}{BC} = \dfrac{AF}{AC}$.

44. Two lines **AOB, POQ** intersect at **O**; the circles **AOP, BOQ** cut again at **X**; prove that $\dfrac{XA}{XP} = \dfrac{XB}{XQ}$.

45. Prove that the common tangents of two non-intersecting circles divide (internally and externally) the line joining the centres in the ratio of the radii.

46. **M** is the mid-point of **AB**; **AXB, MYB** are equilateral triangles on opposite sides of **AB**; **XY** cuts **AB** at **Z**; prove $AZ = 2ZB$.

47. **AB** is a diameter of a circle **ABP**; **PT** is the perpendicular from **P** to the tangent at **A**; prove $\dfrac{PT}{PA} = \dfrac{AP}{AB}$.

48. **APB, AQB** are two circles; if **PAQ** is a straight line, prove that $\dfrac{BP}{BQ}$ equals the ratio of their diameters.

49. **ABCD** is a parallelogram; any line through **C** cuts **AB** produced, **AD** produced at **X**, **Y**; prove $\dfrac{AD}{BX} = \dfrac{DY}{AB}$.

50. **ABCD** is a rectangle; two perpendicular lines are drawn; one cuts **AB**, **CD** at **E**, **F**; the other cuts **AD**, **BC** at **G**, **H**; prove $\dfrac{EF}{GH} = \dfrac{BC}{AB}$.

51. In the quadrilateral **ABCD**, $\angle$ **ABC** = $\angle$ **ADC** and $\dfrac{AB}{BC} = \dfrac{CD}{DA}$; prove **AB** = **CD**.

52. The diagonals **AC**, **BD** of the quadrilateral **ABCD** meet at **O**; if the radius of the circle **AOD** is three times the radius of the circle **BOC**, prove **AD** = 3**BC**.

53. **ABCD** is a parallelogram; **P** is any point on **AB**; **DP** cuts **AC** at **Q**; prove $\dfrac{AP}{AB} = \dfrac{PQ}{DQ}$.

54. **AB**, **DC** are the parallel sides of the trapezium **ABCD**; any line parallel to **AB** cuts **CA**, **CB** at **H**, **K**; **DH**, **DK** cut **AB** at **X**, **Y**; prove **AB** = **XY**.

55. **ABCD** is a parallelogram; **O** is any point on **AC**; lines **POQ**, **ROS** are drawn, cutting **AB**, **CD**, **BC**, **AD** at **P**, **Q**, **R**, **S**; prove **PS** is parallel to **QR**.

56. In $\triangle$**ABC**, **D** is the mid-point of **BC**; **AD** is bisected at **E**; **BE** cuts **AC** at **F**; prove **CF** = 2**FA**. [Draw **EK** parallel to **BC** to cut **AC** at **K**.]

57. **BC**, **YZ** are the bases of two similar triangles **ABC**, **XYZ**; **AP**, **XQ** are medians; prove $\angle$ **BAP** = $\angle$ **YXQ**.

58. **P** is a variable point on a given circle; **O** is a fixed point outside the circle; **Q** is a point on **OP** such that **OQ** = $\tfrac{1}{3}$**OP**; prove that the locus of **Q** is a circle.

59. **ABC** is a $\triangle$; **E**, **F** are the mid-points of **AB**, **AC**; **EFD** is drawn so that **FD** = 2**EF**; prove **BF** bisects **AD**.

60. In $\triangle$**ABC**, $\angle$ **BAC** = 90°; **ABXY**, **ACZW** are squares outside $\triangle$**ABC**; **BZ**, **CX** cut **AC**, **AB** at **K**, **H**; prove **AH** = **AK**.

61. In $\triangle$**ABC**, the bisectors of $\angle$s **ABC**, **ACB** meet at **D**; **DE**, **DF** are drawn parallel to **AB**, **AC** to meet **BC** at **E**, **F**; prove $\dfrac{BE}{FC} = \dfrac{BA}{AC}$.

SIMILAR TRIANGLES

62*. In $\triangle ABC$, $\angle BAC = 90°$; AD is an altitude; H, K are the in-centres of $\triangle$s ADB, ADC; prove that $\triangle$s DHK, ABC are similar.

63*. D, E, F are the mid-points of the sides BC, CA, AB of a triangle; O is any other point; prove that the lines through D, E, F parallel to OA, OB, OC are concurrent.

64*. In $\triangle ABC$, $AB = n \cdot AC$; BQ is the perpendicular from B to the bisector of $\angle BAC$; BC cuts AQ at P; prove that $\dfrac{PQ}{PA} = \dfrac{n-1}{2}$.

RECTANGLE PROPERTIES OF A CIRCLE

THEOREM 56

(i) If two chords **AB** and **CD** of a circle intersect at a point **O** (inside or outside a circle),
then **OA . OB = OC . OD**.

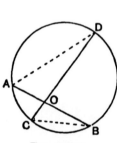
Fig. 129(1).

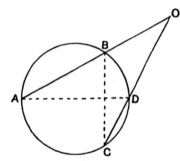
Fig. 129(2).

(ii) If from any point **O** outside a circle, a line is drawn *touching* the circle at **T**, and another line is drawn cutting the circle at **A, B**,
then **OA . OB = OT²**.

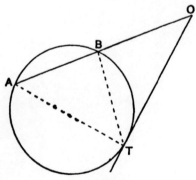
Fig. 130.

RECTANGLE PROPERTIES OF A CIRCLE   121

THEOREM 57

If **AD** is an altitude of the triangle **ABC**, which is right-angled at **A**,

then   (i) $AD^2 = BD \cdot DC$;   (ii) $BA^2 = BD \cdot BC$.

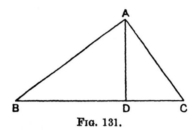

FIG. 131.

DEFINITION.—If a, x, b are such that $\dfrac{a}{x} = \dfrac{x}{b}$ or $x^2 = ab$,

x is called the *mean proportional* between a and b.

The converse properties are important:—

  (i) If two lines **AOB, COD** are such that $AO \cdot OB = CO \cdot OD$, then **A, B, C, D** lie on a circle.

  (ii) If two lines **OAB, ODC** are such that $OA \cdot OB = OC \cdot OD$, then **A, B, C, D** lie on a circle.

  (iii) If two lines **OBA, OT** are such that $OA \cdot OB = OT^2$, then the circle through **A, B, T** touches **OT** at **T**.

Alternative proof of Theorem 57:—

  (i) Draw the circle on **BC** as diameter: it passes through **A**, since $\angle BAC = 90°$. Produce **AD** to cut the circle again at **E**.

  Since the chord **AE** is perp. to diameter **BC**, $AD = DE$.

  But $AD \cdot DE = BD \cdot DC$;

  $\therefore \quad AD^2 = BD \cdot DC$.

  (ii) Draw the circle on **AC** as diameter: it passes through **D**, since $\angle ADC = 90°$, and touches **BA** at **A**, since $\angle BAC = 90°$.

  $\therefore$ by Theorem 56 (ii), $BA^2 = BD \cdot BC$.

RECTANGLE PROPERTIES OF A CIRCLE

EXERCISE XXIII

1. Find a mean proportional between (i) 3 and 48 ; (ii) $12x$, $3xy^2$.
2. From a point **P** on a circle, **PN** is drawn perpendicular to a diameter **AB** ; **AN** = 3″, **NB** = 12″ ; find **PN**.
3. In △**ABC**, ∠**BAC** = 90° ; **AD** is an altitude ; **AB** = 5″, **AC** = 12″ ; find **BD**.
4. In △**ABC**, **AB** = 8, **AC** = 12 ; a circle through **B**, **C** cuts **AB**, **AC** at **P,Q** ; **BP** = 5 ; find **CQ**.
5. The diagonals of a cyclic quadrilateral **ABCD** meet at **O** ; **AC** = 9, **BD** = 12, **OA** = 4 ; find **OB**.
6. In Fig. 132,
    (i) If **AB** = 9, **BO** = 3, find **OT**.
    (ii) If **OB** = 6, **OT** = 12, find **AB**.
    (iii) If **OA** = 3, **AB** = 2, **AT** = 4, find **BT**.
    (iv) If **AB** = 8, **AT** = 6, **BT** = 5, find **OT**.

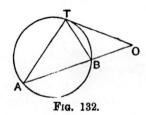

Fig. 132.

7. **ABC** is a triangle inscribed in a circle ; **AB** = **AC** = 10″ **BC** = 12″ ; **AD** is drawn perpendicular to **BC** and is produced to meet the circle in **E** ; find **DE** and the radius of the circle.
8. In △**ABC**, ∠**ABC** = 90°, **AB** = 3″, **BC** = 4″ ; find the radius of the circle which passes through **A** and touches **BC** at **C**.
9. In △**ABC**, ∠**BAC** = 90° ; **AD** is an altitude ; **BC** = a, **CA** = b, **AB** = c, **AD** = h, **BD** = x, **DC** = y ; prove that (i) $h^2 = xy$; (ii) $b^2 = y(x+y)$; (iii) $hc = bx$; (iv) $\dfrac{b^2}{c^2} = \dfrac{y}{x}$.
10. In Fig. 132, if **OA** = 2**OT**, prove **AB** = 3**BO**.

11. **AOB, COD** are two perpendicular chords of a circle, centre **K**; **AO** = 6, **CO** = 10, **OD** = 12; find **OK, AK**.
12. **X** is the mid-point of a line **TY** of length 2″; **TZ** is drawn so that $\angle$**ZTX** = 45°; a circle is drawn through **X, Y** touching **TZ** at **P**; prove $\angle$**TXP** = 90°, and find the radius of the circle.
13. **ABC** is a $\triangle$ inscribed in a circle; the tangent at **C** meets **AB** produced in **D**; **BC** = p, **CA** = q, **AB** = r, **BD** = x, **CD** = y; find x, y in terms of p, q, r.
14. Express, in the form of equal ratios, the equations: (i) $xy = ab$; (ii) $pq = r^2$; (iii) **OA . OB = OC . OD**; (iv) **ON . OT = OP^2**.
15. The diagonals of a cyclic quadrilateral **ABCD** intersect at **O**; prove **AD . OC = BC . OD**.
16. Two lines **OAB, OCD** cut a circle at **A, B, C, D**; prove **OA . BC = OC . AD**.
17. Two chords **AB, CD** of a circle intersect at **O**; if **D** is the mid-point of arc **AB**, prove **CA . CB = CO . CD**.
18. In $\triangle$**ABC**, **AB = AC** and $\angle$**BAC** = 36°; the bisector of $\angle$**ABC** meets **AC** at **P**; prove **AC . CP = BC^2 = AP^2**.
19. The altitudes **BE, CF** of $\triangle$**ABC** intersect at **H**; prove that (i) **BH . HE = CH . HF**; (ii) **AF . AB = AE . AC**; (iii) **CE . CA = CH . CF**.
20. In $\triangle$**ABC**, **AB = AC**; **D** is a point on **AC** such that **BD = BC**; prove **BC^2 = AC . CD**.
21. Two circles intersect at **A, B**; **P** is any point on **AB** produced; prove that the tangents from **P** to the circles are equal.
22. In $\triangle$**ABC**, $\angle$**BAC** = 90°, **AB = 2AC**; **AD** is an altitude; prove **BD = 4DC**.
23. **PQ** is a chord of a circle, centre **O**; the tangents at **P, Q** meet at **T**; **OT** cuts **PQ** at **N**; prove **ON . OT = OP^2**.
24. **AB** is a diameter of a circle; **PQ** is a chord; the tangent at **B** meets **AP, AQ** at **X, Y**; prove **AP . AX = AQ . AY**.
25. **AB, AC** are two chords of a circle; any line parallel to the tangent at **A** cuts **AB, AC** at **D, E**; prove **AB . AD = AE . AC**.
26. **ABCD** is a cyclic quadrilateral; **P** is a point on **BD** such that $\angle$**PAD** = $\angle$**BAC**; prove that (i) **BC . AD = AC . DP**; (ii) **AB . CD = AC . BP**; (iii) **BC . AD + AB . CD = AC . BD**.

27. **AB** is a diameter of a circle, centre **O**; **AP, PQ** are equal chords; prove **AP . PB = AQ . OP**.
28. **AD** is an altitude of $\triangle$ **ABC**; prove that the radius of the circle **ABC** equals $\dfrac{AB \cdot AC}{2AD}$. [Draw diameter through **A**.]
29. Two circles intersect at **A, B**; **PQ** is their common tangent; prove **AB** bisects **PQ**.
30. In $\triangle$**ABC**, **AC** is equal to the diagonal of the square described on **AB**; **D** is the mid-point of **AC**; prove $\angle$ **ABD** = $\angle$ **ACB**.
31. A line **PQ** is divided at **R** so that $PR^2 = PQ \cdot RQ$; **TQR** is a $\triangle$ such that **TQ = TR = PR**; prove **PT = PQ**.
32. **PQR** is a $\triangle$ inscribed in a circle; the tangent at **P** meets **QR** produced at **T**; prove $\dfrac{TQ}{TR} = \dfrac{PQ^2}{PR^2}$.
33. In $\triangle$**ABC**, $\angle$ **BAC** = 90°; **E** is a point on **BC** such that **AE = AB**; prove $BE \cdot BC = 2AE^2$.
34. **AD** is an altitude of $\triangle$**ABC**; if $AB \cdot BC = AC^2$ and if **AB = CD**, prove $\angle$ **BAC** = 90°.
35. Two chords **AB, AC** of a circle are produced to **P, Q** so that **AB = BP** and **AC = CQ**; if **PQ** cuts the circle at **R**, prove $AR^2 = PR \cdot RQ$.
36. The tangent at a point **C** on a circle is parallel to a chord **DE** and cuts two other chords **PD, PE** at **A, B**; prove $\dfrac{AC}{CB} = \dfrac{AD}{BE}$.
37. **AB** is a diameter of a circle, centre **O**; the tangents at **A, B** meet any other tangent at **H, K**; prove $AH \cdot BK = AO^2$.
38. Two lines **OAB, OCD** cut a circle at **A, B, C, D**; through **O**, a line is drawn parallel to **BC** to meet **DA** produced at **X**; prove $XO^2 = XA \cdot XD$.
39. **ABC** is a $\triangle$ inscribed in a circle; a line through **B** parallel to **AC** cuts the tangent at **A** in **P**; a line through **C** parallel to **AB** cuts **AP** in **Q**; prove $\dfrac{AP}{AQ} = \dfrac{AB^2}{AC^2}$.
40*. **AB** is a chord of a circle **APB**; the tangents at **A, B** meet at **T**; **PH, PK, PX** are the perpendiculars to **TA, TB, AB**; prove $PH \cdot PK = PX^2$.

RECTANGLE PROPERTIES OF A CIRCLE

41*. **AB, AC** are tangents to the circle **BDCE**; **ADE** is a straight line; prove **BE.CD = BD.CE**.

42*. **P, Q** are points on the radius **OA** and **OA** produced of a circle, centre **O**, such that **OP.OQ = OA^2**; **R** is any other point on the circle; prove that **RA** bisects $\angle$ **PRQ**.

43*. In $\triangle$**ABC**, **AB = AC**, $\angle$ **BAC = 36°**; prove **AB^2 − BC^2 = AB.BC**.

44*. The internal bisector of $\angle$ **BAC** cuts **BC** at **D**, prove that **AD^2 = BA.AC − BD.DC**. [Use ex. 17.]

45* The external bisector of $\angle$ **BAC** cuts **BC** produced at **E**; prove that **AE^2 = BE.EC − BA.AC**.

46*. **ABCD** is a parallelogram; **H, K** are fixed points on **AB, AD**; **HP, KQ** are two variable parallel lines cutting **CB, CD** at **P, Q**; prove **BP.DQ** is constant.

AREAS AND VOLUMES

THEOREM 58

If **ABC, XYZ** are two similar triangles, and if **BC, YZ** are a pair of corresponding sides,

then $\dfrac{\triangle ABC}{\triangle XYZ} = \dfrac{BC^2}{YZ^2}.$

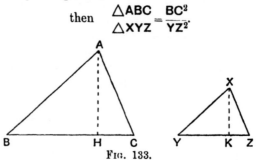

Fig. 133.

More generally, the ratio of the areas of any two similar polygons is equal to the ratio of the squares on corresponding sides.

THEOREM 59

If **AB** and **CD** are corresponding sides of any two similar polygons **PAB, QCD**, and if **AB, CD, EF** are three lines in proportion $\left(i.e.\ \dfrac{AB}{CD} = \dfrac{CD}{EF}\right),$

then $\dfrac{\text{figure PAB}}{\text{figure QCD}} = \dfrac{AB}{EF}.$

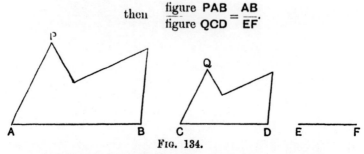

Fig. 134.

The following facts are also of importance (see ex. 34, 35) :—
- (i) The ratio of the areas of the surfaces of similar solids equals the ratio of the squares of their linear dimensions.
- (ii) The ratio of the volumes of similar solids equals the ratio of the cubes of their linear dimensions.

AREAS AND VOLUMES

EXERCISE XXIV

1. A screen, 6' high (not necessarily rectangular) requires 27 sq. ft. of material for covering; how much is needed for a screen of the same shape, 4' high?
2. On a map whose scale is 6" to the mile, a plot of ground is represented by a triangle of area $2\frac{1}{4}$ sq. inches; what is the area (in acres) of the plot?
3. The sides of a triangle are 6 cms., 9 cms., 12 cms.; how many triangles whose sides are 2 cms., 3 cms., 4 cms. can be cut out of it? How would you cut it up?
4. Show how to divide any triangle into 25 triangles similar to it.
5. The area of the top of a table, 3 feet high, is 20 sq. ft.; the area of its shadow on the floor is 45 sq. ft.; find the height of the lamp above the floor.
6. A light is 12 feet above the ground; find the area of the shadow of the top of a table 4 ft. high, 9 ft. long, 5 ft. broad.
7. **ABC**, **XYZ** are similar triangles; **AD**, **XK** are altitudes; **AB** = 15, **BC** = 14, **CA** = 13, **AD** = 12, **XY** = 5; find **XK** and the ratio of the areas of △s **ABC**, **XYZ**.
8. A triangle **ABC** is divided by a line **HK** parallel to **BC** into two parts **AHK**, **HKCB** of areas 9 sq. cms., 16 sq. cms.; **BC** = 7 cms.; find **HK**.
9. **E** is the mid-point of the side **AB** of a square **ABCD**; **AC** cuts **ED** at **O**; **AB** = 3"; find the area of **EBCO**.
10. **ABC** is a △ such that **AB** = **AC** = 2**BC**; **D** is a point on **AC** such that ∠ **DBC** = ∠ **BAC**; a line through **D** parallel to **BC** cuts **AB** in **E**; find the ratio of the areas △**ABC** : △**BCD** : △**BED** : △**EDA**.
11. Water in a supply pipe of diameter 1 ft. comes out through a tap 3" in diameter: in the pipe it is moving at 5" a second; with what velocity does it come out of the tap?
12. If it costs £3 to gild a sphere of radius 3 ft., what will it cost to gild a sphere of radius 4 ft.?
13. Two hot-water cans are the same shape; the smaller is 9" high

128 CONCISE GEOMETRY

and holds a quart; the larger is 15" high: how much will it hold?

14. How many times can a cylindrical tumbler 4" high and 3" in diameter be filled from a cylindrical cask 40" high and 30" in diameter?

15. A metal sphere, radius 3", weighs 8 lb.; find the weight of a sphere of the same metal 1' in radius.

16. A cylindrical tin 5" high holds ¼ lb. of tobacco; how much will a tin of the same shape 8" high hold?

17. Two models of the same statue are made of the same material; one is 3" high and weighs 8 oz.; the other weighs 4 lb.; what is its height?

18. A lodger pays 8 pence for a scuttle of coal, the scuttle being 20" deep; what would he pay if the scuttle was the same shape and 2½ feet deep.

19. A tap can fill half of a spherical vessel, radius 1½ feet, in 2 minutes; how long will two similar taps take to fill one-quarter of a spherical vessel of radius 4 feet?

20. Two leaden cylinders of equal lengths and diameters 3", 4" are melted and recast as a single cylinder of the same length what is its diameter?

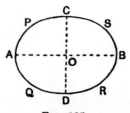

Fig. 135.

21. In the given figure, not drawn to scale, the lines **AB**, **CD** bisect each other at right angles; **AB** = 6 cms., **CD** = 4 cms., **PAQ**, **RBS** are arcs of circles of radii 1 cm.; **PCS**, **QDR** are arcs of circles of radii 3½ cms., touching the former arcs. Construct a similar figure in which the length of the line corresponding to **AB** is 9 cms.

The area of the first figure is approximately 18 sq. cms., what is the area of the enlarged figure?

AREAS AND VOLUMES

If in the given figure, the curve is rotated about **AB** to form an egg-shaped solid, its volume is approximately 48 c.c.; what is the volume of the solid obtained similarly from the enlarged figure?

22. The sides of a $\triangle$**ABC** are trisected as in the figure; prove that the area of **PQRSXY** $= \frac{2}{3} \triangle$**ABC**.

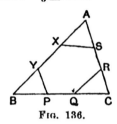

Fig. 136.

23. If in the $\triangle$s **ABC, XYZ**, $\angle$ **BAC** = $\angle$ **YXZ**, prove that $\dfrac{\triangle \mathbf{ABC}}{\triangle \mathbf{XYZ}} = \dfrac{\mathbf{AB} \cdot \mathbf{AC}}{\mathbf{XY} \cdot \mathbf{XZ}}$.

24. Two lines **OAB, OCD** meet a circle at **A, B, C, D**, prove that $\dfrac{\triangle \mathbf{OAD}}{\triangle \mathbf{OBC}} = \dfrac{\mathbf{AD}^2}{\mathbf{BC}^2}$. What result is obtained by making **B** coincide with **A**?

25. **H, K** are any points on the sides **AB, AC** of $\triangle$**ABC**, prove that $\dfrac{\triangle \mathbf{AHK}}{\triangle \mathbf{ABC}} = \dfrac{\mathbf{AH} \cdot \mathbf{AK}}{\mathbf{AB} \cdot \mathbf{AC}}$.

26. In $\triangle$**ABC**, $\angle$ **BAC** $= 90°$ and **AD** is an altitude; prove $\dfrac{\mathbf{AB}^2}{\mathbf{AC}^2} = \dfrac{\mathbf{BD}}{\mathbf{DC}}$.

27. **ABCD** is a parallelogram; **P, Q** are the mid-points of **CB, CD**; prove $\triangle$ **APQ** $= \frac{3}{8}$ parallelogram **ABCD**.

28. Any circles through **B, C** cuts **AB, AC** at **D, E**; prove $\dfrac{\triangle \mathbf{ADE}}{\triangle \mathbf{ABC}} = \dfrac{\mathbf{DE}^2}{\mathbf{BC}^2}$.

29. In $\triangle$**ABC**, $\angle$ **BAC** $= 90°$ and **AD** is an altitude; **DE** is the perpendicular from **D** to **AB**; prove $\dfrac{\mathbf{BE}}{\mathbf{BA}} = \dfrac{\mathbf{BA}^2}{\mathbf{BC}^2}$.

30. **AP** is a chord and **AB** is a diameter of a circle, centre **O**; the tangents at **A, P** meet at **T**; prove $\dfrac{\triangle \mathbf{TAP}}{\triangle \mathbf{POB}} = \dfrac{\mathbf{AP}^2}{\mathbf{PB}^2}$.

31. **ABC** is an equilateral triangle; **BC** is produced each way to **P, Q**; if $\angle PAQ = 120°$, prove $\dfrac{PB}{CQ} = \dfrac{AP^2}{AQ^2}$.

32. In $\triangle ABC$, $\angle BAC = 90°$; **BCX, CAY, ABZ** are similar triangles with **X, Y, Z** corresponding points; prove $\triangle CAY + \triangle ABZ = \triangle BCX$.

33. A room is lighted by a single electric bulb in the ceiling; a table with level top is moved about in the room; prove that the area of the shadow of the top on the floor does not alter.

34. If x ins. is the length of some definite dimension in a figure of given shape, its area $= kx^2$ sq. ins. where k is constant for different sizes. Find k for (i) square, side x; (ii) square, diagonal x; (iii) circle, radius x; (iv) circle, perimeter x; (v) equilateral triangle, side x; (vi) regular hexagon, side x; (vii) surface of cube, side x; (viii) surface of sphere, radius x.

35. If x ins. is the length of some definite dimension in a figure of given shape, its volume $= kx^3$ cu. ins. where k is constant for different sizes. Find k for (i) cube, edge x; (ii) cube, diagonal x; (iii) sphere, diameter x; (iv) sphere, equator x; (v) the greatest circular cylinder that can be cut from a cube, edge x; (vi) circular cone, vertical angle $90°$, height x; (vii) regular tetrahedron, edge x.

THE BISECTOR OF THE VERTICAL ANGLE OF A TRIANGLE

Theorem 60

(i) **ABC** is a triangle; if the line bisecting $\angle$ **BAC** (internally or externally) cuts **BC**, or **BC** produced at **D**,

then $\dfrac{BA}{AC} = \dfrac{BD}{DC}$.

(ii) If **D** is a point on the base **BC**, or **BC** produced of the triangle **ABC** such that $\dfrac{BA}{AC} = \dfrac{BD}{DC}$, then **AD** bisects internally or externally $\angle$ **BAC**.

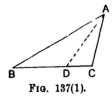

Fig. 137(1).   Fig. 137(2).

THE BISECTOR OF THE VERTICAL ANGLE OF A TRIANGLE

EXERCISE XXV

1. In $\triangle ABC$, $AB = 6$ cms., $BC = 5$ cms., $CA = 4$ cms.; the internal and external bisectors of $\angle BAC$ cut BC and BC produced at P, Q; find BP and BQ and show that $\dfrac{1}{BP} + \dfrac{1}{BQ} = \dfrac{2}{BC}$.

2. In $\triangle ABC$, $AB = 4''$, $BC = 3''$, $CA = 5''$; the bisector of $\angle ACB$ cuts AB at D; find CD.

3. In $\triangle ABC$, $AB = 12$, $BC = 15$, $CA = 8$; P is a point on BC such that $BP = 9$; prove AP bisects $\angle BAC$; if the external bisector of $\angle BAC$ cuts BC produced at Q, and if D is the mid-point of BC, prove that $DP \cdot DQ = DC^2$.

4. The internal and external bisectors of $\angle BAC$ meet BC and BC produced at P, Q; $BP = 5$, $PC = 3$; find CQ.

5. $ABCD$ is a rectangular sheet of paper; $AB = 4''$, $BC = 3''$; the edge BC is folded along BD and the corner is then cut off along the crease; find the area of the remainder.

6. In $\triangle ABC$, $AB = 6''$, $AC = 4''$; the bisector of $\angle BAC$ meets the median BE at O; the area of $\triangle ABC$ is 8 sq. in.; what is the area of $\triangle AOB$?

---

7. The internal and external bisectors of $\angle BAC$ cut BC and BC produced at P, Q; prove $\dfrac{BP}{PC} = \dfrac{BQ}{CQ}$.

8. AX is a median of $\triangle ABC$; the bisectors of $\angle$s AXB, AXC meet AB, AC at H, K; prove HK is parallel to BC.

9. $ABCD$ is a parallelogram; the bisector of $\angle BAD$ meets BD at K; the bisector of $\angle ABC$ meets AC at L; prove LK is parallel to AB.

BISECTOR OF VERTICAL ANGLE OF A TRIANGLE 133

10. The tangent at a point A of a circle, centre O, meets a radius OB at T; D is the foot of the perpendicular from A to OB; prove $\dfrac{DB}{BT} = \dfrac{AD}{AT}$.

11. The bisector of $\angle$ BAC cuts BC at D; circles with B, C as centres are drawn through D and cut BA, CA at H, K; prove HK is parallel to BC.

12. H is any point inside the $\triangle$ABC; the bisectors of $\angle$s BHC, CHA, AHB cut BC, CA, AB at X, Y, Z; prove $\dfrac{BX}{XC} \times \dfrac{CY}{YA} \times \dfrac{AZ}{ZB} = 1$.

13. Two lines OAB, OCD meet a circle at A, B, C, D; the bisector of $\angle$ AOC cuts AC, BD at H, K; prove $\dfrac{AH}{HC} = \dfrac{DK}{KB}$.

14. The bisector of $\angle$ BAC cuts BC at D; the circle through A, B, D cuts AC at P; the circle through A, C, D cuts AB at Q; prove BQ = CP.

15. Two circles, centres A, B, touch at O; any line parallel to AB cuts the circles at P, Q respectively; AP and BQ are produced to meet at K; prove OK bisects $\angle$ AKB.

16. A straight line cuts four lines OP, OQ, OR, OS at P, Q, R, S; if $\angle$ POR = 90° and OR bisects $\angle$ QOS, prove $\dfrac{PQ}{PS} = \dfrac{QR}{RS}$.

17. The tangent at a point T on a circle cuts a chord PQ when produced at O; the bisector of $\angle$ TOP meets TP, TQ at X, Y; prove $TX^2 = TY^2 = PX \cdot QY$.

18. In $\triangle$ABC, $\angle$ BAC = 90° and AD is an altitude; the bisector of $\angle$ ABC meets AD, AC at L, K; prove $\dfrac{AL}{LD} = \dfrac{CK}{KA}$.

19. ABCD is a quadrilateral; if the bisectors of $\angle$s DAB, DCB meet on DB, prove that the bisectors of $\angle$s ABC, ADC meet on AC.

20. Two circles touch internally at O; a chord PQ of the larger touches the smaller at R; prove $\dfrac{OP}{OQ} = \dfrac{PR}{RQ}$.

21*. If I is the in-centre of $\triangle$ABC, and if AI meets BC at D, prove that $\dfrac{AI}{ID} = \dfrac{AB + AC}{BC}$.

22*. The internal and external bisectors of $\angle$ APB meet AB at X, Y; prove $\angle$ XPY $= 90°$. If A, B are fixed points and if P varies so that $\dfrac{PA}{PB}$ is constant, prove that the locus of P is a circle. [*Apollonius'* circle.]

23*. If the internal and external bisectors of $\angle$ BAC meet BC and BC produced at D, E, prove $DE^2 = EB . EC - DB . DC$.

24*. ABC is a triangle such that $AB + AC = 2BC$; the bisector of $\angle$ BAC meets BC at D; prove $AD^2 = 3BD . DC$.

EXAMPLES ON THE CONSTRUCTIONS OF BOOK I

USE OF INSTRUMENTS

EXERCISE XXVI

USE OF RULER, DIVIDERS, AND PROTRACTOR

1. Measure in inches and cms. the lines a, b, c, d.

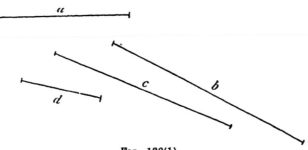

FIG. 138(1).

2. Draw a straight line across your sheet of paper and mark off by eye lengths of 4 cms., 7 cms., 2 inches; then measure them and write down your errors.
3. Draw a line and cut off from it a length of 5"; measure it in cms. and find the number of cms. in 1 inch.
4. Draw a line and cut off from it a length of 10 cms.; measure it in inches and hence express 1 cm. in inches.
5. In Fig. 138(2), measure in cms. the lengths of **AC, BD, BC, AD**. What are the values of (i) **AC + BD**; (ii) **AD + BC**.

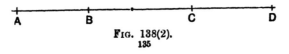

FIG. 138(2).

6. Measure in inches and cms. the length of this page. Taking 1" = 2·54 cms. approx., find how far your measurements agree with each other.
7. Draw a straight line across your paper: mark the middle point by eye and measure the two parts. How far is the point you have marked from the real mid-point of the line?
8. Draw a straight line across your paper and divide it by eye into three equal parts: measure the three parts.
9. Repeat ex. 8, dividing the line into four equal parts.
10. Draw a straight line across your paper and use your dividers (i) to bisect it; (ii) to trisect it.
11. It is required to obtain points on a line **AB** produced beyond an obstacle which obstructs the view. **C** is one of the points required, perform the construction and verify it.

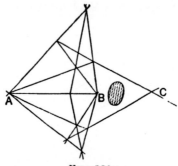

Fig. 139.

12. Measure the angles a, b, c, d.

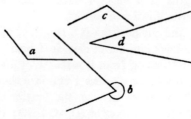

Fig. 140.

13. Use your protractor to draw angles of (i) 30°, (ii) 90°, (iii) 48°, (iv) 124°, (v) 220°, (vi) 300°.

USE OF INSTRUMENTS 137

14. Measure the angles *a, b, c, d* and write down their sum.

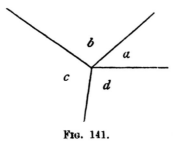

Fig. 141.

15. Measure the angles *a, b* and write down their sum.

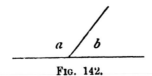

Fig. 142.

16 Measure the angles *a, x, b, y*. What do you notice about them?

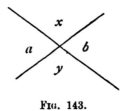

Fig. 143.

17. Measure the angles AOB, BOC, AOC.

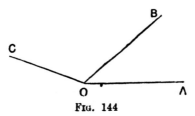

Fig. 144

18. Measure the three angles of the triangle **ABC** and write down their sum.

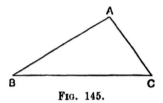

FIG. 145.

19. Measure the three angles of the triangle **DEF** and write down their sum.

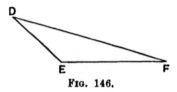

FIG. 146.

20. Without measurement, say which is the larger of the angles, *a* in Fig. 147 or *b* in Fig. 148, and roughly by how much.

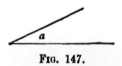

FIG. 147.

21. Draw by eye (with a ruler) angles of 15°, 30°, 60°, 110°, 160°. Measure them and write down your errors.
22. Without measurement state whether the angles *a*, *b*, *c*, *d*, *e* in Fig. 148 are acute or obtuse or reflex.

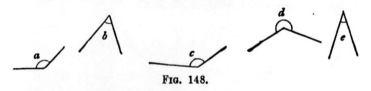

FIG. 148.

23. Draw an angle **ABC** equal to 108°; produce **CB** to **D**. Measure ∠**ABD**.

USE OF INSTRUMENTS

24. Draw an angle **AOB** equal to 82°; produce **AO, BO,** to **C, D**. Measure ∠ **COD**.
25. Draw any five-sided figure **ABCDE** and produce **AB, BC, CD, DE, EA**. Measure each of the five exterior angles so formed and write down their sum.
26. Draw any triangle **ABC**; produce **BC** to **D**. Measure ∠**CBA**, ∠**CAB**, ∠**ACD**. Is ∠**CBA** + ∠**CAB** equal to ∠**ACD**?
27. Draw a figure like Fig. 149; find by measurement the values of ∠**ABC** + ∠**ADC** + ∠**BAD** and ∠**BCD**.

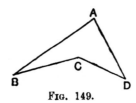

FIG. 149.

28. Enlarge Fig. 150, making **AB** = 8 cms., **AD** = **BC** = 2 cms., ∠**DAB** = 90° = ∠**CBA**. **O** is the mid-point of **AB**. Mark points **F, G, H, K, L, M, N** on **CD** such that the lines joining them to **O** make with **OB** angles of 30°, 50°, 70°, 90°, 110°, 130°, 150°. Measure in cms. **FG, GH, HK**.

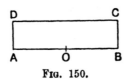

FIG. 150.

USE OF COMPASSES

29. Draw a circle, centre **O**; draw any diameter **AB**; take any three points **P, Q, R** on the circumference. Measure ∠s **APB, AQB, ARB**.
30. Draw two circles of radii 3 cms., 4 cms. so that their centres are 5 cms. apart. Draw their common chord, *i.e.* the line

joining the points at which they cut, and measure its length. What is the angle at which it cuts the line joining the centres?

31. Take two points A, B 3 cms. apart; construct two points P, Q such that PA = PB = 5 cms. = QA = QB.
32. Take a point P; describe a circle of radius 4 cms. passing through P; construct a chord PQ of length 6 cms.
33. Draw a circle; take four points A, B, C, D in order on it. Measure (i) ∠ACB and ∠ADB; (ii) ∠ABC and ∠ADC. What do you notice?
34. Draw a large triangle ABC (not isosceles); describe circles on AB and AC as diameters. Do they meet on BC?
35. Take two points A, B 5 cms. apart. Construct a point C such that CA = 6 cms., CB = 7 cms. Draw circles with centres A, B, C and radii 2, 3, 4 cms. respectively. What do you notice about them?
36. Take two points A, B 3 cms. apart. Construct a point C such that CA = CB = 6 cms. Join CA, CB and measure ∠CAB, ∠CBA, ∠ACB. Is ∠CAB equal to ∠CBA? Is ∠CAB equal to twice ∠ACB?
37. Draw a circle of radius 3 cms. and place in it 6 chords each of length 3 cms., end to end; what figure is obtained? Measure the angle between two adjacent chords.
38. Draw a straight line AB; construct a point C such that CA = CB = AB. Measure the angles of △ABC.
39. Draw a straight line AB and take any point P outside it. Construct a point Q such that QA = PA and QB = PB. Join PQ and let it cut AB at R. Measure ∠PRA.
40. Draw two circles of radii 3 cms., 4 cms. so that the part of the line joining their centres which lies inside both circles is 1 cm.
41. Draw a line AB 5 cms. long; construct a point C so that CA = 3 cms., CB = 4 cms. Join CA, CB. Bisect with dividers or by measurement AB at D. Measure ∠ACB and CD. Is CD = $\frac{1}{2}$AB?
42. Draw a line AB 3 cms. long; construct a circle of radius 4 cms. to pass through A and B.

USE OF INSTRUMENTS 141

43. Take two points **A, B** 6 cms. apart. Construct 10 positions of a point **P** (on either side of **AB**) such that **PA + PB** = 10 cms. (*e.g.* **PA** = 3, **PB** = 7 or **PA** = 4, **PB** = 6, etc.). All these positions lie on a smooth curve called an *ellipse*: draw freehand a curve through these positions. Would you expect the curve to pass through **A** or **B**?

44. Draw a circle, centre **O**, and take any point **T** outside it; on **TO** as diameter describe a circle cutting the first at **P, Q**. Join **TP, TQ** and produce both. What do you notice about these lines?

45. Draw a circle, centre **O**, of radius 3·5 cms.; draw a chord **PQ** such that $\angle$ **POQ** = 72°. Construct four other chords **QR, RS**, etc., end to end, each equal to **PQ**. What is the figure so obtained?

46. Draw two unequal circles intersecting at **P, Q**; draw the diameters **PX, PY** of the circles. Join **XY**. Does **XY** pass through **Q**?

47. Draw a circle, centre **O**, and take any six points **A, B, C, D, E, F** in order on the circumference. Measure $\angle$ s **ABF, ACF, ADF, AEF, AOF**. Do you notice any connection between them?

48. Draw any angle **AOB**; with **O** as centre and any radius (not too short), describe a circle cutting **OA, OB** at **P, Q**; with **P, Q** as centres and any radius (not too short), describe two equal circles cutting at **R**. Measure $\angle$ **AOR**, $\angle$ **BOR**.

    This construction enables you *to bisect a given angle*.

49. Draw any straight line **AB**; with **A, B** as centres and any radius (not too short), describe two equal circles cutting at **P, Q**. Join **PQ** and let it cut **AB** at **R**. Measure **AR, RB** and $\angle$ **ARP**.

    This construction enables you *to draw the perpendicular bisector of a given straight line.*

50. Draw any straight line **AB** and take any point **C** on it.

    With **C** as centre, describe any circle cutting **AB** at **P, Q**; with **P, Q** as centres and any radius (not too short), describe two equal circles cutting at **R**. Join **CR**. Measure $\angle$ **ACR**.

This construction enables you *to draw a straight line perpendicular to a given straight line from a given point on the line.*

51. Draw any straight line **AB** and take any point **C** outside it. With **C** as centre, describe any circle cutting **AB** at **P, Q**; with **P, Q** as centres and any radius (not too short), describe two equal circles cutting at **R**. Join **CR** and let it cut **AB** at **S**. Measure ∠ **ASC**.

    This construction enables you *to draw a straight line perpendicular to a given straight line from a given point outside the line.*

52. Draw any straight line **AB** and take any point **C** outside it. Take any point **P** on **AB**. Join **CP** and bisect it at **Q**. With **Q** as centre and **QC** as radius, describe a circle, cutting **AB** at **R**. Join **CR**. Measure ∠ **ARC**.

    This construction gives an alternate method to Ex. 51.

53. With any point **O** as centre, describe a circle; draw any chord **PQ**: construct the perpendicular bisector of **PQ**. Does it pass through **O**?

54. Draw a triangle **ABC** (not isosceles); construct the perpendicular bisectors of **AB** and **AC**; let them meet at **O**; with **O** as centre and **OA** as radius, describe a circle. Does the circle pass through **B** and **C**?

55. In Fig. 151, without producing **AB**, construct a line through **C** perpendicular to **AB**.

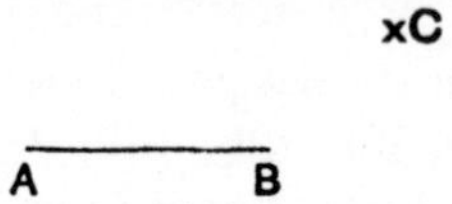

×C

A   B

Fig. 151.

56. Draw a line **AB**, construct a line through **B** perpendicular to **AB** *without* producing **AB**.

57. Draw an obtuse-angled triangle **ABC**; construct the perpendiculars from each vertex to the opposite side. Are they concurrent?

58. Draw a circle and take four points **A, B, C, X** on it; construct the perpendiculars **XP, XQ, XR** to **BC, CA, AB**. What do you notice about **P, Q, R**?

USE OF INSTRUMENTS

59. Draw a circle of radius 3 cms. and take points **A, B, C** on it such that **AB** = 4 cms., **AC** = 5 cms. Measure ∠ **BAC**: is there more than one answer?

60. Draw a line **AB** and take any two points **C, D** outside it; construct a point **P** on **AB** such that **PC** = **PD**.

61. Draw any triangle (not isosceles) and construct the bisectors of its three angles. What do you notice about them?

62. Draw any triangle **ABC**; construct the bisectors of ∠s **ABC, ACB** and let them meet at **I**. Construct the perpendicular **IX** from **I** to **BC**. With **I** as centre and **IX** as radius, describe a circle. What do you notice about this circle?

63. Draw two lines **ABC, BD**, cutting at **B**; construct the bisectors **BP, BQ** of ∠ **ABD**, ∠ **CBD**; measure ∠ **PBQ**.

64. Construct (without using a protractor) angles of (i) 30°, (ii) 45°, (iii) 105°, (iv) 255°.

65. Draw a circle and take any three points **A, B, C** on it (**AB** ≠ **AC**); construct the perpendicular bisector of **BC** and the bisector of ∠ **BAC** and produce them to meet. What do you notice about their point of intersection?

66. Draw an obtuse angle and construct lines dividing it into four equal angles.

67. Draw a triangle **ABC** (not isosceles); construct a point **P** on **BC** such that the perpendiculars from **P** to **AB** and **AC** are equal.

68. Draw a right angle and construct the lines trisecting it.

69. Draw a line **PQ** (see Fig. 152), cutting two other lines **AB, CD** at **P, Q**; the bisectors of ∠s **APQ, CQP** meet at **H**; the bisectors of ∠s **BPQ, DQP** meet at **K**; verify that **HK** when produced passes through the point of intersection of **AB** and **CD** and bisects the angle between them.

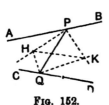

Fig. 152.

144　CONCISE GEOMETRY

70. Copy the following figures 153–167 on any convenient scale.

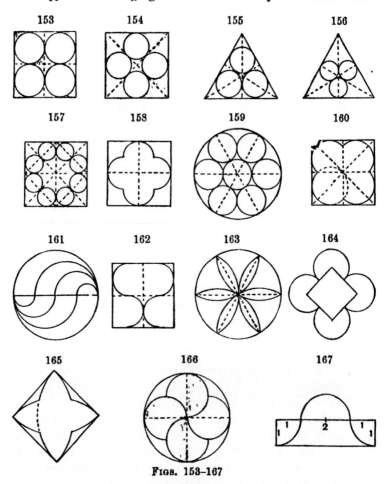

Figs. 153–167

USE OF SET SQUARES

71. Draw a line **AB** and take three points **P, Q, R** outside it: use set squares to draw lines through **P, Q, R** parallel to **AB**.

72. Draw a line **AB** and take three points **P, Q, R** outside it: use set squares to draw lines through **P, Q, R** perpendicular to **AB**.

DRAWING TO SCALE

73. Draw a line **AB** and take a point **C** on it: use set squares to draw a line through **C** perpendicular to **AB**.
74. Draw a line **AB** and take a point **P** outside it: use set squares to draw two lines **PQ**, **PR** making angles of 60° with **AB**.
75. Draw a triangle **ABC** and use set squares to draw its three altitudes (*i.e.* perpendiculars from each corner to the opposite side).
76. Draw a triangle **ABC** and use set squares to complete the parallelogram **ABCD**: measure its sides.
77. Use set squares to draw a four-sided figure having its opposite sides parallel and one angle a right angle: measure the diagonals.
78. Draw a triangle **ABC** (not isosceles); bisect **AB** at **H**; use set squares to draw a line **HK** parallel to **BC** to meet **AC** at **K**; measure **AK**, **KC**.
79. Draw any angle **BAC** and cut off **AB** equal to **AC**; use set squares to construct bisector of ∠ **BAC**.
80. Use set squares to draw a right angle, and use them to trisect it.
81. Draw a line **AB** and divide it into five equal parts as follows: draw any other line **AC** and cut off from **AC** five equal parts **AP**, **PQ**, **QR**, **RS**, **ST**; join **BT**; through **P**, **Q**, **R**, **S** draw lines parallel to **TB**: these cut **AB** at the required points.

DRAWING TO SCALE

EXERCISE XXVII

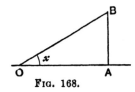

Fig. 168.

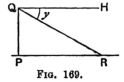

Fig. 169.

DEFINITIONS.—(i) In Fig. 168, if **OA** is horizontal, ∠ **AOB** is called *the angle of elevation* of **B** as viewed from **O**.

(ii) In Fig. 169, if **QH** is horizontal, ∠ **HQR** is called *the angle of depression* of **R** as viewed from **Q**.

1. A courtyard is 80 feet long and 50 feet wide; what is the distance between two opposite corners?

2. A gun whose range is 5000 yards is in position at a point 3500 yards from a straight railway line; what length of the line can it command?

3. A ladder, 15 feet long, is resting against a vertical wall; the foot of the ladder is 6 feet from the wall; how high up the wall does it reach?

4. The ends of a cord, 10 feet long, are fastened to two nails each of which is 15 feet above the ground; the nails are 5 feet apart; a weight is attached to the mid-point of the cord: how high is it above the ground?

5. A straight passage runs from A to B, then turns through an angle of 70° and runs on to C; if AB is 80 yards and BC is 100 yards, what distance is saved by having a passage direct from A to C?

6. A man rows due north at 4 miles an hour, and the current takes him north-east at 5 miles an hour; how far is he from his starting-point after 20 minutes?

7. A man starts from A and walks 2 miles due south to B, then 3 miles south-west to C, then 1 mile west to D; what is the direction and distance of D from A?

8. Southampton is 12 miles S.S.W. of Winchester; Romsey is 10 miles W. 32° S. of Winchester. Find the distance and bearing of Romsey from Southampton.

9. An aeroplane points due north and flies at 60 miles an hour; the wind carries it S.W. at 15 miles an hour. What is its position ten minutes after leaving the aerodrome?

10. Andover is 12 miles from Winchester and 15 miles from Salisbury; Salisbury is 20 miles W. of Winchester. [Andover is north of the Salisbury-Winchester line.] Find the bearing of Andover from Salisbury.

11. Exeter is 42 miles from Dorchester and 64 miles from Bristol; Bristol is 55 miles due north of Dorchester; Barnstable is 33 miles N.E. of Exeter. What is the distance and bearing of Barnstable from Dorchester?

12. A weight is slung by two ropes of lengths 12 feet, 16 feet, from two pegs 18 feet apart in a horizontal line. What is the depth of the weight below the line of the pegs?

13. From two points 500 yards apart on a straight road running due north, the bearings of a house are found to be N. 40° E. and E. 20° S.; find the shortest distance of the house from the road.
14. There are two paths inclined at an angle of 40° which lead from a gate across a circular field: one runs across the centre of the field and is 120 yards long; what is the length of the other?
15. A path runs round the edge of a square ploughed field **ABCD**; if you follow the path from **A** to **C** you go 50 yards farther than if you walk straight across. What is the length of a side of the field?
16. One end of a string, 5 feet long, is fastened to a nail, and a weight is attached to the other end; the weight swings backwards and forwards through 15° each side of the vertical. What is the distance between its two extreme positions?
17. At a distance of 40 yards from a tower, the angle of elevation of the top of the tower is 35°; find the height of the tower in feet.
18. A kite is flown at the end of a string 120 yards long which makes an angle of 65° with the ground: find in feet the height of the kite.
19. What is the elevation of the sun when a pole 12 feet high casts a shadow 20 feet long?
20. A fenced level road running due north suddenly turns due east, with the result that the shadow of the fence is increased in breadth from 3 feet to 5 feet: what is the bearing of the sun?
21. The elevation of the top of a chimney is 20°; from a place 60 yards nearer, it is 30°; find its height in feet.
22. From the top of a cliff 150 feet high, the angle of depression of a boat out at sea is 20°; what is the distance of the boat from the cliff in yards?
23. From the top of a tower 250 feet high, the angles of depression of two houses in a line with and at the same level as the foot of the tower are $6\frac{1}{4}$° and 48°. Find their distance apart in yards.

MISCELLANEOUS CONSTRUCTIONS—I

EXERCISE XXVIII

1. Draw an angle **BAC** and a line **PQ**; construct points **R, S** on **AB, AC** such that **RS** is equal and parallel to **PQ**.
2. Draw a circle and construct points **P, Q, R** on it such that **PQ = QR = RP**; take any other point **X** on the circle. Measure **XP, XQ, XR** and verify that the longest of these equals the sum of the other two.
3. Draw an angle **BAC** of 50°; construct on **AB, AC** points **P, Q** such that $\angle$ **QPA** = 90° and **PQ** = 4 cms. Measure **AP**.
4. Draw a circle of radius 4 cms., and take a point **A** at a distance of 2·5 cms. from the centre: construct a chord **PQ** passing through **A** and bisected at **A**.
5. Draw a large quadrilateral **ABCD**, so that **AB** is not parallel to **CD**; construct a point **P** such that **PA = PB** and **PC = PD**.
6. Draw a line **AB** and take a point **C** distant 2″ from **AB**; construct a circle with **C** as centre, cutting **AB** at two points 3″ apart. Measure its radius.
7. Draw an angle **BAC** of 70°; construct a point **P** whose distances from **AB, AC** are 3 cms., 4 cms. Measure **AP**.
8. Draw a line **AB** and take a point **C** distant 2″ from **AB**; construct two points **P, Q** each of which is 1½″ from **AB** and 1½″ from **C**. Measure **PQ**.
9. Draw two lines **AB, AC** and take a point **P** somewhere between them; construct a line to pass through **P** and cut off equal lengths from **AB** and **AC**.
10. Draw two lines **AB, CD** and take any point **E** between them. Construct a line to pass through **E** and the (inaccessible) point of intersection of **AB, CD**. [Use the system of parallel lines shown in Fig. 170.]

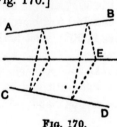

Fig. 170.

MISCELLANEOUS CONSTRUCTIONS—1

11. Draw a triangle **ABC**; construct a line through **C** parallel to the bisector of ∠**BAC** and let it meet **BA** produced at **E**. Measure **AE**, **AC**.
12. Draw a circle and take two points **A**, **B** outside it. Construct a circle to pass through **A**, **B** and have its centre on the first circle. When is this impossible?
13. Draw a circle and take a point **H** outside it; draw two lines **HAB**, **HDC**, cutting the circle at **A**, **B**, **D**, **C**; join **AD**, **BC**, and produce them to meet at **K**. Construct a circle to pass through **H**, **A**, **D** and a second circle to pass through **K**, **D**, **C**. Do these circles cut again at a point on **HK**?
14. Construct five points in the same relative position to each other as are **A**, **B**, **C**, **D**, **E** in Fig. 171.

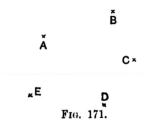

Fig. 171.

15. Take a line **AB** and a point **C** outside it such that the foot of the perpendicular from **C** to **AB** would be off the page. Construct that portion of the perpendicular which comes on the page.
16. Take a line **AB** and a point **C** and suppose there is an obstacle between **C** and **AB** which a set square cannot move over (see Fig. 172). Construct a line through **C** parallel to **AB**.

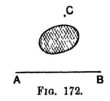

Fig. 172.

17. By folding, obtain a crease which (i) bisects a given angle, (ii) bisects a given line at right angles.

18. By folding, obtain the perpendicular to a given line from a given point outside it.
19. By folding, obtain an angle of 45°.
20. Take a triangular sheet of paper and find by folding the point which is equidistant from the three corners.

CONSTRUCTION OF TRIANGLES, PARALLELOGRAMS, Etc.

EXERCISE XXIX

1. Construct, *when possible*, the triangle **ABC** from the following measurements, choosing your own unit. If there are two different solutions, construct both :—

Fig. 173.

  (i) $a=3$, $b=4$, $c=5$, measure **A**.
 (ii) $a=3$, $b=4$, $c=8$, measure **A**.
(iii) $a=5$, **B**$=30°$, **C**$=45°$, measure b.
 (iv) $a=4$, **A**$=48°$, **B**$=33°$, measure b.
  (v) $a=7$, **A**$=110°$, **B**$=40°$, measure b.
 (vi) $a=5$, **B**$=125°$, **C**$=70°$, measure b.
(vii) $b=5$, $c=7$, **C**$=72°$, measure a.
(viii) $b=6$, $c=4$, **C**$=40°$, measure a.
 (ix) $b=8$, $c=6$, **C**$=65°$, measure a.
  (x) **A**$=40°$, **B**$=60°$, **C**$=80°$, measure a.
 (xi) **A**$=50°$, **B**$=40°$, **C**$=70°$, measure a.
(xii) **A**$=125°$, $b=7\cdot3$, $c=5\cdot4$, measure a.
(xiii) **A**$=90°$, $a=11\cdot2$, $b=7\cdot3$, measure c.
(xiv) $a=b=6\cdot9$, **A**$=50°$, measure c.
 (xv) $a=2b$, $c=\dfrac{3b}{2}$, measure **A**.

2. Draw two unequal lines **AC**, **BD** bisecting each other; join **AB**, **BC**, **CD**, **DA** and measure them. **ABCD** is a *parallelogram*.

CONSTRUCTION OF TRIANGLES

3. Draw two equal lines **AC, BD** bisecting each other; join **AB, BC, CD, DA**; measure ∠**ABC**. **ABCD** is a *rectangle*.
4. Draw two unequal lines **AC, BD** bisecting each other at right angles; join **AB, BC, CD, DA** and measure them. **ABCD** is a *rhombus*.
5. Draw two equal lines **AC, BD** bisecting each other at right angles; join **AB, BC, CD, DA**; measure **AB, BC,** ∠**ABC**. **ABCD** is a *square*.
6. Draw two unequal perpendicular lines **AC, BD** such that **AC** bisects **BD**; join **AB, BC, CD, DA** and measure them. **ABCD** is a *kite*.
7. Draw an angle of 57° and cut off **AB, AC** from the arms of the angle so that **AB** = 5 cms., **AC** = 8 cms.; construct a point **D** such that **BD** = **AC** and **CD** = **AB**. What sort of a quadrilateral is **ABCD** ?
8. Construct a parallelogram **ABCD**, given **AB** = 7 cms., **AC** = 10 cms., **BD** = 8 cms.; measure **BC, CD**.
9. Construct an isosceles triangle with a base of 6 cms. and a vertical angle of 70°; measure its sides.
10. Construct a rhombus **ABCD**, given **AB** = 5 cms., **AC** = 6 cms.; measure ∠**BAD**.
11. Construct an isosceles triangle of base 4·6 cms. and height 5 cms.; measure its vertical angle.
12. Construct the quadrilateral **ABCD**, given **AB** = **BC** = 3 cms., **AD** = **DC** = 5 cms., ∠**ABC** = 120°; measure ∠**ADC**.
13. Construct the rhombus **ABCD**, given **AC** = 6 cms., **BD** = 9 cms.; measure **AB**.
14. Construct the rhombus **ABCD**, given ∠**ABC** = 40°, **BD** = 7 cms.; measure **AC**.
15. Construct a rectangle **ABCD**, given **BD** = 8 cms. and that **AC** makes an angle of 54° with **BD**; measure **AB, BC**.
16. Construct a trapezium **ABCD** with **AB, CD** its parallel sides such that **AB** = 8, **BC** = 4, **CD** = 3, **AD** = 2; measure ∠**BAD**.
17. Construct the quadrilateral **ABCD**, given that
    (i) **AB** = 4, **BC** = 4·5, **CD** = 3, ∠**ABC** = 80°, ∠**BCD** = 110°; measure **AD**:
    (ii) **AB** = 5, **AC** = 6, **AD** = 4, **BD** = 7, **CD** = 3; measure **BC**.

  (iii) $\angle$ **ABC** $= 70°$, $\angle$ **BCD** $= 95°$, $\angle$ **CDA** $= 105°$,   **AB** $= 5$, **AD** $= 4$; measure **BC**.

  (iv) **AB** $= 5$,   **BC** $= 6$,   **CD** $= 3$,   **DA** $= 4·5$, $\angle$ **ADC** $= 100$; measure $\angle$ **ABC**.

  (v) **AB** $= 5$, $\angle$ **CAB** $= 35°$, $\angle$ **ABD** $= 47°$, $\angle$ **ACB** $= 65°$, $\angle$ **ADB** $= 54°$; measure **CD**.

18. Construct the triangle **ABC**, given that
    (i) $a + b = 11$, $b + c = 16$, $c + a = 13$; measure **A**.
    (ii) $A - B = 25°$, $C = 55°$, $c = 7$; measure a.
    (iii) $A : B : C = 1 : 2 : 3$, $a = 3$; measure c.
    (iv) $A + B = 118°$, $B + C = 96°$, $a = 7$; measure c.

19. Construct an equilateral triangle **ABC** such that if **D** is a point on **BC** given by **BD** $= 3$ cms., then $\angle$ **DAC** $= 40°$; measure **BC**.

20. Construct a square having one diagonal 5 cms.; measure its side.

21. **AD** is an altitude of the triangle **ABC**; given **AD** $= 4$ cms., $\angle$ **ABC** $= 55°$, $\angle$ **ACB** $= 65°$, construct $\triangle$ **ABC**; measure **BC**.

22. **AE** is a median of the triangle **ABC**; given **AB** $= 4$ cms., **AC** $= 7$ cms., **AE** $= 4·5$ cms., construct $\triangle$ **ABC**; measure **BC**.

23. **AD** is an altitude of the triangle **ABC**; given **AB** $= 6$ cms. **AD** $= 4$ cms., $\angle$ **ACB** $= 68°$, construct $\triangle$ **ABC**; measure **BC**.

24. **AD** is an altitude of $\triangle$ **ABC**; **AD** $= 4$ cms., $\angle$ **BAC** $= 75°$, $\angle$ **ABC** $= 50°$, construct $\triangle$ **ABC**; measure **BC**.

25. The distances between the opposite sides of a parallelogram are 3 cms., 4 cms., and one angle is $70°$; construct the parallelogram and measure one of the longer sides.

26. Construct a parallelogram of height 4 cms., having its diagonals 5 cms., 8 cms. in length : measure one of the longer sides.

27. Construct an equilateral triangle of height 4 cms.; measure its side.

28. Construct the triangle **ABC**, given that
    (i) $a + b = 2c = 14$, $A = 70°$; measure a.
    (ii) $a + b + c = 20$, $A = 65°$, $B = 70°$; measure a.
    (iii) $a = 10$, $b + c = 13$, $A = 80°$; measure b.
    (iv) $a = 8$, $b + c = 10$, $B = 35°$; measure b.
    (v) $a = 9$, $c - b = 4$, $B = 25°$; measure c.
    (vi) $a = 9$, $b - c = 2$, $A = 70°$; measure b.
    (vii) $a = 5$, $b = 3$, $A - B = 20°$; measure c.

MISCELLANEOUS CONSTRUCTIONS—II

29. Construct an isosceles triangle of height 5 cms. and perimeter 18 cms.; measure its base.
30. Each of the base angles of an isosceles triangle exceeds the vertical angle by 24°; the base is 4 cms.; construct the triangle and measure its other sides.

MISCELLANEOUS CONSTRUCTIONS—II

EXERCISE XXX

1. Given two points **H**, **K** on the same side of a given line **AB**, construct a point **P** on **AB** such that **PH**, **PK** make equal angles with **AB**.
2. Given two points **H**, **K** on opposite sides of a given line **CD**, (see Fig. 174), construct a point **P** on **CD** such that ∠ **HPC** = ∠ **KPC**.

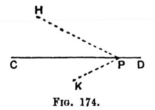

Fig. 174.

3. Given a triangle **ABC**, construct a line passing through **A** from which **B** and **C** are equidistant.
4. Given a triangle **ABC**, construct a line parallel to **BC**, cutting **AB**, **AC** at **H**, **K** such that **BH** + **CK** = **HK**.
5. Given a square **ABCD**, construct points **P**, **Q** on **BC**, **CD** such that **APQ** is an equilateral triangle.
6. Given a triangle **ABC**, construct a rhombus with two sides along **AB**, **AC** and one vertex on **BC**.
7. Given two parallel lines **AB**, **CD** and a point **P** between them, construct a line through **P**, cutting **AB**, **CD** at **Q**, **R** such that **QR** is of given length.
8. Given a triangle **ABC**, construct a point which is equidistant from **B** and **C** and also equidistant from the lines **AB** and **AC**.

9. Given in position the internal bisectors of the angles of a triangle and the position of one vertex, construct the triangle.
10. By construction and measurement, find the height of a regular tetrahedron, each edge of which is 2″.
11. A room is 20 feet long, 15 feet wide, 10 feet high; a cord is stretched from one corner of the floor to the opposite corner of the ceiling, find by drawing and measurement the angle which the cord makes with the floor.
12. Construct a square such that the length of its diagonal exceeds the length of its side by a given length.

EXAMPLES ON THE CONSTRUCTIONS OF BOOK II

AREAS

EXERCISE XXXI

1. Find the areas of the following figures, making any necessary constructions and measurements :—
    (i) △ABC, given $b=5$, $c=4$, $A=90°$.
    (ii) Rectangle ABCD, given $AB=7$, $AC=10$.
    (iii) △ABC, given $a=5$, $b=6$, $c=7$.
    (iv) △ABC, given $b=5$, $c=4$, $B=90°$.
    (v) △ABC, given $b=c=10$, $a=12$.
    (vi) △ABC, given $a=6$, $B=130°$, $C=20°$.
    (vii) ‖gram ABCD, given $AB=8$, $AD=6$, $\angle ABC=70°$.
    (viii) A rhombus whose diagonals are 7, 8.
    (ix) A trapezium ABCD, given $AB=5$, $BC=6$, $CD=9$, $\angle BCD=30°$, and AB parallel to DC.
    (x) Quad. ABCD, given $AB=3$, $BC=5$, $CD=6$, $DA=4$, $BD=5$.
2. Draw a triangle whose sides are 5, 6, 8 cms. and obtain its area in three different ways.
3. Draw a triangle with sides 5, 6, 7 cms., and construct an isosceles triangle with base 6 cms. equal in area to it; measure its sides.
4. Construct a parallelogram of area 21 sq. cms. such that one side is 6 cms., one angle is 50°; measure the other side.
5. Construct a parallelogram of area 15 sq. cms. with sides 5 cms., 6 cms.; measure its acute angle.
6 Draw a triangle with sides 4, 5, 6 cms., and construct a parallelogram equal in area to it and having one side equal

to 4 cms. and one angle equal to 70°; measure the other side.

7. Construct a rhombus each side of which is 5 cms. and of area 15 sq. cms.; measure its acute angle.

8. Draw a parallelogram with sides 4 cms., 6 cms., and one angle 70°; construct a parallelogram of equal area with sides 5 cms., 7 cms.; measure its acute angle.

9. Construct a parallelogram of area 20 sq. cms., with one side 5 cms., and one diagonal 7 cms.; measure the other side.

10. Draw a triangle with sides 5, 6, 8 cms., and construct a triangle of equal area with sides 5·5, 6·5 cms.; measure the third side.

11. Construct a parallelogram equal in area to a given rectangle and having its sides of given length.

12. Construct a triangle equal in area to a given triangle and having one side equal in length to a given line, and one angle adjacent to that side equal to a given angle.

13. Draw a quadrilateral **ABCD** such that **AB** = 6 cms., **BC** = 5 cms., **CD** = 4 cms., ∠**ABC** = 110°, ∠**BCD** = 95°. Reduce it to an equivalent triangle with **AB** as base and its vertex on **BC**. Find its area.

14. Draw a figure like Fig. 149, and reduce it to an equivalent triangle having **AB** as base and its vertex on **AD**.

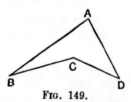

Fig. 149.

15. Draw a figure like Fig. 175 and reduce it to an equivalent triangle.

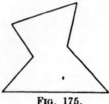

Fig. 175.

AREAS

16. Given four points **A, B, C, D** as in Fig. 176, construct a point **P** such that the figures **ABPD** and **ABCD** are of equal area and **DP** is perpendicular to **AB**.

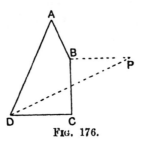

Fig. 176.

17. Given a parallelogram **ABCD** and a point **O** inside it, construct a line through **O** which divides **ABCD** into two parts of equal area.
18. Given a triangle **ABC** and a point **D** on **BC** such that $BD < \frac{2}{3}BC$, construct a point **P** on **AC** such that (i) $\triangle DPC = \frac{1}{3}\triangle ABC$, (ii) $\triangle DPC = \frac{2}{7}\triangle ABC$.
19. Given a parallelogram **ABCD**, construct point **P, Q** on **BC, CD** such that **AP, AQ** divide the parallelogram into three portions of equal area.
20. Given a quadrilateral **ABCD**, construct a line through **A** which divides the quadrilateral into two parts of equal area.
21. Given a quadrilateral, construct lines through one vertex which divide it into five parts of equal area.
22. If **ABCD** is any parallelogram, and if **P** is any point on **BD**, and if lines are drawn through **P** parallel to **AB, BC** as in Fig. 177, the parallelograms **AP, PC** are of equal area. Use this fact for the following construction:—
Construct a parallelogram equal in area to and equiangular to a given parallelogram and having one side of given length.

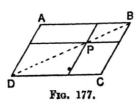

Fig. 177.

23. Given a triangle **ABC**, construct a point **G** inside it such that the triangles **GAB, GBC, GCA** are of equal area.
24. Given a quadrilateral **ABCD**, perform the following construction for a line **BP** bisecting it (see Fig. 178). Bisect **AC** at **O**; through **O** draw **OP** parallel to **BD** to meet **CD** (or **AD**) at **P**; join **BP**.

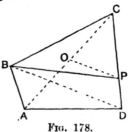

Fig. 178.

SUBDIVISION OF A LINE

EXERCISE XXXII

1. Draw a line **AB**; divide into three equal parts without measuring it.
2. Draw a line **AB** and construct a point **P** on **AB** such that $\dfrac{AP}{PB} = \dfrac{2}{3}$.
3. Draw a line **AB** and construct a point **Q** on **AB** produced, such that $\dfrac{AQ}{BQ} = \dfrac{7}{4}$.
4. Divide a given line in the ratio 5 : 3 both internally and externally.
5. Construct a diagonal scale which can be used for measuring lengths to $\frac{1}{100}$ inch.
6. By using a diagonal scale, draw a line of length 2·73 inches: on this line as base construct an isosceles right-angled triangle and measure its equal sides as accurately as possible.
7. Use a diagonal scale to measure the hypotenuse of a right-angled triangle whose sides are 2″ and 3″.
8. On a scale of 6″ to the mile, what length represents 2000 yards? Draw a scale showing hundreds of yards.

9. What is the R.F. [i.e. *representative fraction*] for a map of scale 2″ to the mile? Construct a scale for reading off distances up to 5000 yards, and as small as 500 yards
10. The R.F. of a map is 1 : 20,000 ; express this in inches to the mile and construct a suitable scale to read miles and tenths of miles.
11. Given two lines **AB**, **AC** and a point **P** between them, construct a line through **P**, cutting **AB**, **AC** at **Q**, **R** so that **QP** = **PR**.
12. Given two lines **AB**, **AC** and a point **P** between them, construct a line through **P** with its extremitie on **AB**, **AC** and divided at **P** in the ratio 2 : 3.
13. Draw a triangle **ABC** such that **BC** = 6 cms.; construct a line parallel to **BC**, cutting **AB**, **AC** at **H**, **K** such that **HK** = 2 cms. What is the ratio **AH** : **HB** ?
14. Given a triangle **ABC**, construct a line parallel to **BC**, cutting **AB**, **AC** at **H**, **K** such that **HK** = $\frac{2}{3}$**BC**.

EXAMPLES ON THE CONSTRUCTIONS OF BOOK III

CONSTRUCTION OF CIRCLES, Etc.

EXERCISE XXXIII

1. Use a coin to draw a circle, and construct its centre.
2. Given two points **A**, **B** and a line **CD**, construct a circle to pass through **A** and **B** and have its centre on **CD**.
3. Draw a line **AB** 3 cms. long, and construct a circle of radius 5 cms. to pass through **A** and **B**.
4. Draw two lines **AOB**, **COD** intersecting at an angle of 80°; make **AO** = 3 cms., **OB** = 4 cms., **CO** = 5 cms., **OD** = 2·4 cms.; construct a circle to pass through **A**, **B**, **C**. Does it pass through **D**?
5. Construct two circles of radii 4 cms., 5 cms., such that their common chord is of length 6 cms. Measure the distance between their centres.
6. Draw two lines **OAB**, **OCD** intersecting at an angle of 40°; make **OA** = 2 cms., **OB** = 6 cms., **OC** = 3 cms., **OD** = 4 cms.; construct a circle to pass through **A**, **B**, **C**. Does it pass through **D**?
7. Given a circle and two points **A**, **B** inside it, construct a circle to pass through **A** and **B** and have its centre on the given circle.
8. Given a point **B** on a given line **ABC** and a point **D** outside the line, construct a circle to pass through **D** and to touch **AC** at **B**.
9. Draw a line **AB** and take a point **C** at a distance of 3 cms. from the line **AB**; construct a circle of radius 4 cms. to pass through **C** and touch **AB**.
10. Draw two lines **AB**, **AC** making an angle of 65° with each other; construct a circle of radius 3 cms. to touch **AB** and **AC**.

11. Draw a circle of radius 3 cms. and take a point **A** at a distance of 4 cms. from its centre; construct a circle to touch the first circle and to pass through **A**, and to have a radius of 2 cms. Is there more than one such circle?
12. Given a straight line and a circle, construct a circle of given radius to touch both the straight line and the circle. Is this always possible? If not, state the conditions under which it is impossible.
13. Draw a line **AB** of length 6 cms.: with **A**, **B** as centres and radii 3 cms., 2 cms. respectively, describe circles. Construct a circle to touch each of these circles and have a radius of 5 cms. Give all possible solutions. [The contacts may be internal or external.]
14. Draw a circle of radius 4·5 cms., and draw a diameter **AB**; construct a circle of radius 1·5 cm. to touch the circle and **AB**.
15. Given a circle and a point **A** on the circle and a point **B** outside the circle, construct a circle to pass through **B** and to touch the given circle at **A**.
16. Draw a circle of radius 5 cms.; construct two circles of radii 1·5 cm., 2·5 cms. touching each other externally and touching the first circle internally.
17. Draw a triangle whose sides are of lengths 2, 3, 4 cms., and construct the four circles which touch the sides of this triangle and measure their radii.
18. Draw two lines **OA**, **OB** such that $\angle AOB = 40°$, and **OA** = 4 cms.; construct a circle touching **OA** at **A** and touching **OB**; measure its radius.
19. Given a triangle **ABC**, construct a circle to touch **AB**, **AC** and have its centre on **BC**. Is there more than one solution?
20. Inscribe a circle in a given sector of a circle. [*i.e.* Given two radii **OA**, **OB** of a circle, construct a circle to touch **OA**, **OB** and the arc **AB**.]
21. Given two radii **OA**, **OB** of a circle, construct points **H**, **K** on **OA**, **OB** such that the circle on **HK** as diameter touches the arc **AB**.
22. Given two points **A**, **B** and a point **D** on a line **CDE**, construct two concentric circles one of which passes through **A**, **B** and the other touches **CE** at **D**. When is this impossible?

CONSTRUCTION OF CIRCLES, ETC.

23. Given three points **A, B, C**, construct three circles with these points as centres so that each circle touches the other two. Is there more than one solution?
24. Draw two lines **OA, OB** intersecting at an angle of 46°; construct a circle touching **OA** and **OB** and such that the chord of contact is of length 4 cms.; measure its radius.
25. Inscribe a circle in a given rhombus
26. Given two points **A, B**, 4 cms. apart, construct a circle to pass through **A** and **B** and such that the tangents at **A** and **B** include an angle of 100; measure its radius.
27. Find by measurement the radius of the circle inscribed in the triangle whose sides are of lengths 6, 7, 8 cms.
28. **ABC** is a triangle such that **BC** = 6 cms., **BA** = 4 cms., ∠ **ABC** = 90°; find by measurement the radius of the circle escribed to **BC**.
29. Given two parallel lines and a point between them, construct a circle to touch the given lines and pass through the given point.
30. Draw a quadrilateral so that its sides in order are 4, 5, 7, 6 cms.; inscribe a circle in it to touch three of the sides. Does it touch the fourth side?
31. In Fig. 179, **AB, CD** are two given parallel lines: construct a circle to touch **AB, CD** and the given circle.

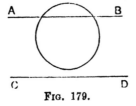

FIG. 179.

32. Given two parallel lines **AB, CD** and a circle between them, construct a circle to touch **AB, CD** and to touch and enclose the given circle.
33. Given two circles, centres **A, B**, radii a, b, and a point **C** on the first, construct a circle to touch the first circle at **C** and also to touch the second. Fig. 180 gives the construction for the centre **P** of the required circle, if it touches both circles externally. **D** is found by making **CD** = b. Perform

11 this construction and construct also the circle in the case where the contacts are external with circle **A**, internal with circle **B**. How would **C** be situated if the constructed circle touches circle **A** internally and circle **B** either internally or externally?

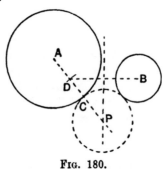

Fig. 180.

34. **ABC** is an equilateral triangle; **AB** = 4 cms.; **A**, **B** are the centres of two equal circles of radii 2·5 cms.; **CA** is *produced* to meet the first circle at **D**. Construct a circle touching the first circle internally at **D** and touching the second circle externally. State your construction.

35. Construct a circle to touch a given line **AB** and a given circle centre **C**, at a given point **D**. Fig. 181 gives the construction for the centre **P** of the required circle if the contact is external. Perform the construction and construct the case where the contact is internal.

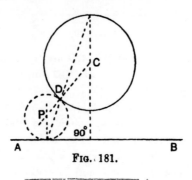

Fig. 181.

CONSTRUCTION OF CIRCLES, ETC. 165

Construct the Figs. in exs. 36–62 : **do not rub out any of your construction lines.**

36. Three arcs each of radius 3 cms. and each $\frac{1}{6}$th of a complete circumference.

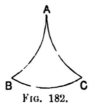

FIG. 182.

37. **AB, BC, CD, DE** are equal quadrants ; **AE** = 6 cms.

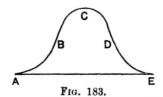

FIG. 183.

38. **AB, BC, CD, DE, EF, FG, GH, HA** are alternately semicircles and quadrants of equal radius ; **XY** = 10 cms.

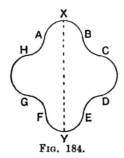

FIG. 184.

39. Three arcs each of radius 3 cms. touching at **A, B, C**.

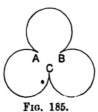

FIG. 185.

40. The sides of the rectangle are 6 cms., 8 cms.

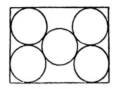

Fig. 186.

41. The radii of the arcs **AB**, **BC**, **CA** are 3·5 cms., 2·5 cms., 7 cms.

Fig. 187.

42. The radii of the circles are 1 cm., 2 cms., 2 cms., 3 cms., and the centre of the smallest circle lies on the largest.

Fig. 188.

43. **AP**, **AQ** are arcs of radii 4 cms.; **PQ** is of radius 8 cms.; **AB** is perpendicular to **CD** and equals 3 cms.

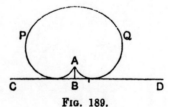

Fig. 189.

CONSTRUCTION OF CIRCLES, ETC.

44. **AB, BC, CD** are arcs of radii 3 cms., **AD** equals 7 cms. and touches **AB, DC**.

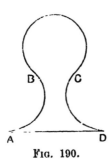

Fig. 190.

45. The radii of the arcs **AB, BC** are 3·5 cms., 1·2 cm., **CD** = 5 cms., **DE** = 6·5 cms., **AE** = 7 cms.

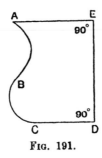

Fig. 191.

46. **AB, AD** are arcs of radii 6 cms.; **AC** equals 6 cms. and is an axis of symmetry.

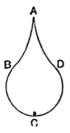

Fig. 192.

47. **BC** is a quadrant; the radii of arcs **AB**, **BC**, **CA** are 4, 2, 3 cms.

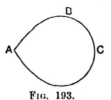
Fig. 193.

48. **AF** is an axis of symmetry; **AB**, **BC**, **DE** are equal quadrants; **AF** = 8 cms., **EG** = 6 cms.

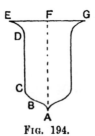
Fig. 194.

49. The radii of the arcs **ABC**, **ADC** are 3, 5·5 cms.; chord **AC** = 5 cms. Construct the figure and inscribe in it a circle of radius 1·5 cm.

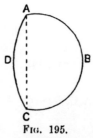
Fig. 195.

50. **CE** is an axis of symmetry; **AB**, **BC** are arcs each of radius 3 cms.; the centre of **AB** lies on **AD**. **AD** = 10 cms., **CE** = 5 cms.

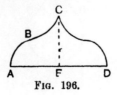
Fig. 196.

CONSTRUCTION OF CIRCLES, ETC.

51. **AB** is an arc of radius 3 cms.; **BC, CD, DA** are arcs each of radius 1 cm.; chord **AE** = chord **EB** = 3 cms.

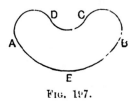

Fig. 197.

52. **AB, BC** are semicircles, each of radius 2 cms. intersecting at an angle of 120°. The arc **AC** touches arcs **AB, CB**.

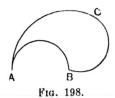

Fig. 198.

53. **AB, DE** are arcs each of radius 2 cms. with their centres on **AE**; **BC** = **CD** = 4 cms.; **AE** = 6 cms.

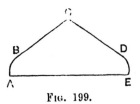

Fig. 199.

54. **CD** is an axis of symmetry; **AB** = 9·5 cms., **CD** = 3·5 cms.; **AE, EC** are arcs of radii 2, 10·5 cms. respectively.

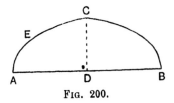

Fig. 200.

55. **AB** is a quadrant of radius 2·5 cms. with its centre on **AC**; **AC** = 7 cms. The arc **BC** touches **AB** at **B**.

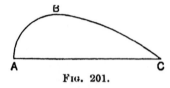

Fig. 201.

56. **ABCD** is a square of side 2 cms.; **BE**, **EF** are circular arcs with **C**, **A** as centres respectively.

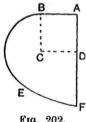

Fig. 202.

57. **AB** is an axis of symmetry; **PAQ** is a semicircle of radius 2 cms.; **RBS** is an arc of radius 1 cm.; **AB** = 7 cms. The arcs **PR**, **QS** are tangential at each end.

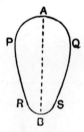

Fig. 203.

58. **AB** = 3·5 cms., **AC** = 6 cms., $\angle$ **BAC** = 90°; radius of arc **CP** is 1·5 cm.

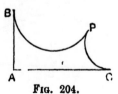

Fig. 204.

CONSTRUCTION OF CIRCLES, ETC.

59. **AB** = **BC** = 3 cms.; the arcs **AB**, **BC** cut the line **ABC** at angles of 30°.

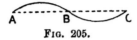

FIG. 205.

60. **AB** is a semicircle, radius 3 cms., centre **O**; **OP**, **OQ** are arcs each of radius 1 cm.; the arcs **AP**, **AB** are tangential at **A**.

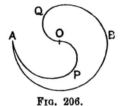

FIG. 206.

61. Fig. 207 is formed by parts of nine equal circles touching where they meet; **AX**, **BY**, **CZ** are each axes of symmetry; the radius of each arc is 1·5 cm.

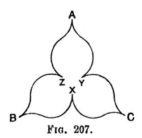

FIG. 207.

62. **AB**, **CD**, **EF**, **GH** are the diameters of semicircles each of radius 1 cm. and when produced form a square; **AD**, **BG**, **CF**, **HE** are arcs each of radius 5 cms.

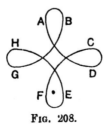

FIG. 208.

MISCELLANEOUS CONSTRUCTIONS—III
EXERCISE XXXIV

1. Draw a circle of radius 3 cms., and construct a chord of the circle of length 5 cms. Take a point **A** inside the circle but not on the chord, and construct a chord of length 5 cms. passing through **A**.
2. Given a chord **PQ** of a given circle and a point **R** on **PQ**, construct a chord through **R** equal to **PQ**.
3. Inscribe a regular hexagon in a given circle.
4. Inscribe an equilateral triangle in a given circle.
5. **A, B, C** are three given points on a given circle; construct a chord of the circle equal to **AB** and parallel to the tangent at **C**.
6. Draw a circle radius 4 cms. and take a point 6 cms. from the centre. Construct the tangents from this point to the circle and measure their lengths.
7. Draw a circle of radius 3 cms., and construct two tangents which include an angle of 100°.
8. Draw a line **AB** of length 7 cms.; construct a line **AP** such that the perpendicular from **B** to **AP** is 5 cms.
9. Draw a circle, centre **O**, radius 4 cms.; take a point **A** 6 cms. from **O**; draw **AB** perpendicular to **AO**; construct a point **P** on **AB** such that the tangent from **P** to the circle is of length 5·5 cms.; measure **AP**.
10. Draw a circle of radius 3 cms. and take a point 5 cms. from the centre; construct a chord of the circle of length 4 cms. which when produced passes through this point.
11. Draw a line **AB** of length 5 cms. and describe a circle with **AB** as diameter; construct a point on **AB** produced such that the tangent from it to the circle is of length 3 cms.
12. Given a circle and a straight line, construct a point on the line such that the tangents from it to the circle contain an angle equal to a given angle.
13. Circumscribe an equilateral triangle about a given circle.
14. On a line of length 5 cms., construct a segment of a circle containing an angle of 70°; measure its radius.

MISCELLANEOUS CONSTRUCTIONS—III 173

15. On a line of length 2 inches, construct a segment of a circle containing an angle of 140°; measure its radius.
16. In a circle of radius 3 cms., inscribe a triangle whose angles are 40°, 65°, 75°; measure its longest side.
17. Inscribe in a circle of radius 1" a rectangle of length 1 5", and measure its breadth.
18. Circumscribe about a circle of radius 2 cms. a triangle whose angles are 50°, 55°, 75°; measure its longest side.
19. Given three non-collinear points A, B, C, construct the tangent at A to the circle which passes through A, B, C *without* either drawing the circle or constructing its centre.
20. Draw two circles of radii 2 cms., 3 cms., with their centres 6·5 cms. apart; construct their four common tangents.
21. Draw two circles of radii 2·5 cms., 3·5 cms., touching each other externally, and construct their exterior common tangents.
22. Draw a line AB of length 6 cms. and construct a line PQ such that the perpendiculars to it from A, B are of lengths 2 cms., 4 cms.
23. Draw two circles of radii 2 cms., 3 cms., with their centres 6 cms. apart; construct a chord of the larger circle of length 4 cms. which when produced touches the smaller circle.
24. Construct the triangle ABC, given that BC = 6 cms., ∠ BAC = 90°, the altitude AD = 2 cms.; measure AB, AC.
25. Construct the triangle ABC, given that BC = 5 cms., ∠ BAC = 55°, the altitude AD = 4 cms.; measure AB, AC.
26. Construct the triangle ABC, given the length of BC and the altitude BE and the angle BAC.
27. Construct a triangle given its base and vertical angle and the length of the median through the vertex.
28. Construct a triangle ABC, given BC = 6 cms., ∠ BAC = 52°, and the median BE = 5 cms.
29. Draw a circle of radius 3 cms., and construct points A, B, C on the circumference such that BC = 5 cms., BA + AC = 8·1 cms.; measure BA and AC.
30. Draw a circle of radius 3·5 cms. and inscribe in it a triangle ABC such that BC = 5·8 cms., BA − AC = 2 cms.; measure BA and AC.

31. Construct a triangle **ABC** given its perimeter, the angle **BAC** and the length of the altitude **AD**.
32. Draw any circle and take two points **A, B** on it and a point **C** outside the circle; construct a point **P** on the circle such that **PC** bisects ∠ **APB**.
33. Draw two lines which meet at a point off your paper; construct the bisector of the angle between them.
34*. Draw any triangle **ABC** (not right-angled). Construct a square **PQRS** such that **PQ** passes through **A**, **QR** passes through **B**, and **PR** cuts **QS** at **C**.
35*. Construct the quadrilateral **ABCD**, given that **AD** = 5 cms., **BC** = 4·6 cms., ∠ **ABD** = ∠ **ACD** = 55°, ∠ **CBD** = 43°; measure **CD**.
36*. Draw any circle and take two points **A, B** on it; construct a point **P** on the circle such that chord **PA** equals twice chord **PB**.
37*. Draw a circle of radius 3 cms., centre **O**, and take a point **P** at distance of 5 cms. from **O**; construct a line through **P**, cutting the circle at **Q, R** such that the segment **QR** contains an angle of 70°; measure ∠ **OPQ**.
38*. Draw two unequal circles intersecting at **A, B**; construct a line through **A**, cutting the circles at **P, Q** such that **PA** = **AQ**.
39*. Draw two unequal circles intersecting at **A, B**; construct a line through **A**, cutting the circles at **P, Q** such that **PQ** is of given length.
40*. Circumscribe a square about a given quadrilateral.

EXAMPLES ON THE CONSTRUCTIONS OF BOOK IV

PROPORTION AND SIMILAR FIGURES

EXERCISE XXXV

1. Construct and measure a fourth proportional to lines of length 4, 5, 6 cms.
2. Construct and measure a third proportional to lines of length 5, 6 cms.
3. Draw a line **AB** and divide it internally in the ratio 2 : 3.
4. Draw a line **AB** and divide it externally (i) in the ratio 5 : 3; (ii) in the ratio 3 : 5.
5. Draw a line **AB** and divide it internally and externally in the ratio 3 : 7.
6. Use a construction to solve $\dfrac{x}{3} = \dfrac{7}{5}$.
7. Find graphically the value of (i) $\dfrac{2 \cdot 3 \times 5 \cdot 9}{4 \cdot 7}$; (ii) $3 \cdot 8 \times 2 \cdot 7$.
8. Construct a line of length $\dfrac{11}{7}$ cms.
9. Draw a line **AB** and divide it in the ratio 2 : 7 : 3.
10. Draw any triangle **ABC** and any line **PQ**; construct a triangle such that its perimeter equals **PQ** and its sides are in the ratio **AB** : **BC** : **CA**.
11. To construct the expressions (i) $\dfrac{ab}{f}$, (ii) $\dfrac{abc}{fg}$, proceed as follows:

    Draw two lines **OH**, **OK** (see Fig. 209).
    From **OH**, cut off **OA** = a.
    From **OK**, cut off **OB** = b, **OC** = c, **OF** = f, **OG** = g.
    Join **AF**, draw **BX** parallel to **FA**, cutting **OH** at **X**, then
    $$\mathbf{OX} = \dfrac{ab}{f}.$$

Join **XG**, draw **CY** parallel to **GX**, cutting **OH** at **Y**, then
$$OY = \frac{ab}{f} \cdot \frac{c}{g} = \frac{abc}{fg}.$$
Use this construction to find (i) $\frac{5 \cdot 1 \times 3 \cdot 8}{4 \cdot 7}$, (ii) $\frac{1 \times 3 \cdot 8 \times 2 \cdot 7}{4 \cdot 7 \times 1 \cdot 8}$

and extend it to find $\frac{abcd}{fgh}$, where a, b, c, d, e, f, g, h are given lengths.

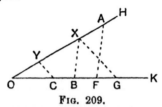

FIG. 209.

12. If a, b, c, d are given numbers, construct, by the method of ex. 11, Fig. 209, (i) $\frac{a}{b}$; (ii) $\frac{1}{ab}$; (iii) $\frac{ab}{cd}$.

13. Given two lines **AB**, **AC** and a point **D** between them, construct a line through **D**, cutting **AB**, **AC** at **P**, **Q** such that $PD = \frac{2}{3}DQ$.

14. Draw a line **ABCD**; if $AB = x$ cms., $BC = y$ cms., $CD = z$ cms., construct a line of length xyz cms.

15. Given a triangle **ABC**, construct a point **P** on **BC** such that the lengths of the perpendiculars from **P** to **AB** and **AC** are in the ratio $2:3$.

16. **ABC** is an equilateral triangle of side 5 cms., construct a point **P** inside it such that the perpendiculars from **P** to **BC**, **CA**, **AB** are in the ratio $1:2:3$. Measure **AP**.

17. Draw any triangle **ABC**, use the method indicated in Fig. 210 to construct a triangle **XYZ** similar to triangle **ABC** and such that $XY = 2AB$.

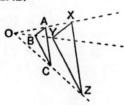

FIG. 210.

PROPORTION AND SIMILAR FIGURES 177

18. Given a quadrilateral **ABCD**, construct a similar quadrilateral each side of which is $\frac{5}{3}$ of the corresponding side of **ABCD**.
19. Given a triangle **ABC** and its median **AD**, construct a similar triangle **XYZ** and its median **XW**, such that $XW = \frac{3}{4}AD$.
20. Construct an equilateral triangle such that the length of the line joining one vertex to a point of trisection of the opposite side is $2''$; measure its side.
21. Using a protractor, construct a regular pentagon such that the perpendicular from one corner to the opposite side is of length 7 cms.; measure its side.
22. Construct a square **ABCD**, given that the length of the line joining **A** to the mid-point of **BC** is $3''$; measure its side.
23. Construct a triangle **ABC**, given $\angle BAC = 54°$, $\angle ABC = 48°$, and the sum of the three medians is 12 cms. Measure **AB**.
24. Inscribe in a given triangle a triangle whose sides are parallel to the sides of another given triangle.
25. Given two radii **OA**, **OB** of a circle, centre **O**; construct a square such that one vertex lies on **OA**, one vertex on **OB**, and the remaining vertices on the arc **AB**.
26. Inscribe a regular octagon in a square.
27. Inscribe in a given triangle **ABC** an equilateral triangle, one side of which is perpendicular to **BC**.
28. Construct a circle to touch two given lines and a given circle, centre **O**, radius a. [Draw two lines parallel to the given lines at a distance a from them: construct a circle to touch these lines and pass through **O**. Its centre is the centre of the required circle.]
29. Draw a line **AB** and take a point **O** $1''$ from it; **P** is a variable point on **AB**; **Q** is a point such that $OQ = OP$ and $\angle POQ = 50°$. Construct the locus of **Q**. [The locus of **Q** is obtained by revolving **AB** about **O** through $50°$.]
30. **ABC** is a given triangle; **P** is a variable point on **BC**; **Q** is a point such that the triangles **ABC**, **APQ** are similar. Construct the locus of **Q**. [Use the idea of ex. 29.]
31. **APQ** is a triangle of given shape; **A** is a fixed point, **P** moves on a fixed circle; construct the locus of **Q**. [Use the idea of ex. 29.]

32*. Given a triangle **ABC** and a point **D** on **BC**, construct points **P, Q** on **AB, AC** such that **DPQ** is an equilateral triangle.
33*. **ABC** is a straight rod whose ends **A, C** move along two perpendicular lines **OX, OY**; **AB** = 6 cms., **BC** = 3 cms. Draw the position of the rod when it makes an angle of 30° with **OX**, and construct the direction in which **B** is moving at this instant.
34*. **AB** and **BC** are two equal rods hinged together at **B**; the end **A** is fixed and **C** is made to move along a fixed line **AX**; **D** is the mid-point of the rod **BC**; construct the direction in which **D** is moving when $\angle$**BAC** = 45°.
35*. A piece of cardboard in the shape of a triangle **ABC** moves so that **AB** and **AC** always touch two given fixed pins **E, F**; draw the triangle in any position and construct the direction in which **A** is moving at that instant.

THE MEAN PROPORTIONAL

EXERCISE XXXVI

1. Construct a mean proportional between 5 and 8; measure it.
2. Construct a line of length $\sqrt{43}$ cms. [Don't take a mean between 1 and 43, this leads to inaccurate drawing; take numbers closer together, such as 5 and 8·6, $\frac{43}{5} = 8·6$.]
3. Find graphically $\sqrt{37}$.
4. Solve graphically the equation $(x-3)^2 = 19$.
5. Draw a rectangle of sides 4 cms., 7 cms., and construct a square of equal area; measure its side.
6. Construct a square equal in area to an equilateral triangle of side 5 cms.; measure its side.
7. Construct a square equal in area to a quadrilateral **ABCD** given **AB** = **BC** = 4, **CD** = 6, **DA** = 7, **AC** = 6 cms.; measure its side.
8. Draw a line **AB**; construct a point **P** on **AB** such that $\dfrac{AP}{AB} = \dfrac{1}{\sqrt{2}}$.
9. Draw a circle, centre **O**; construct a concentric circle whose area is one-third of the first circle.

THE MEAN PROPORTIONAL

10. Given a triangle ABC, construct a line parallel to BC, cutting AB, AC at P, Q so that $\triangle APQ = \frac{1}{2} \triangle ABC$.
11. Given a quadrilateral ABCD, construct a similar quadrilateral with its area $\frac{2}{5}$ of the area of ABCD.
12. Given a triangle ABC, construct an equilateral triangle of equal area.
13. Given three lines whose lengths are a, b, c cms., construct a line of length x cms. such that $\frac{x}{a} = \frac{b^2}{c^2}$.
14. Given two equilateral triangles, construct an equilateral triangle whose area is the sum of their areas.
15. Construct a circle to pass through two given points A, B and touch a given line CD.

    Use the method indicated in Fig. 211 and obtain two solutions.

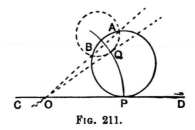

FIG. 211.

16. Given a circle and two points A, B outside it, construct a point P on AB such that PA . PB is equal to the square of the tangent from P to the circle.
17. Construct a circle to pass through two given points A, B and to touch a given circle.
18. Given four points A, B, C, D in order on a straight line, construct a point P on BC such that PA . PB = PC . PD.
19. Solve graphically the equations $x - y = 5$, $xy = 16$.
20. OA, OB are two lines such that OA = 6 cms., $\angle AOB = 40°$; construct a circle touching OA at A and intercepting on OB a length of 5 cms.
21. Construct a circle to pass through a given point, touch a given circle and have its centre on a given line.
22. Given three circles, each external to the others, construct a

point such that the tangents from it to the three circles are of equal length.

23. Draw a circle of radius 5 cms. and take a point **A** 3 cms. from the centre; construct a chord **PQ** of the circle passing through **A** such that **PA** = $\frac{2}{3}$**AQ**.

MISCELLANEOUS CONSTRUCTIONS—IV

EXERCISE XXXVII

1. Draw a line **AB**; if **AB** is of length x inches, construct a line of length x^2 inches.
2. **AB**, **CD** are two given parallel lines, and **O** is any given point; construct a line **OPQ**, cutting **AB**, **CD** at **P**, **Q** so that **AP** : **CQ** is equal to a given ratio.
3. **ABC** is a given equilateral triangle of side 5 cms.; construct a line outside it such that the perpendiculars from **A**, **B**, **C** to the line are in the ratio 2 : 3 : 4 and measure the last.
4. Construct a triangle **ABC**, given ∠**BAC** = 48°, ∠**BCA** = 73°, and the median **BE** = 5 cms.; measure **AC**.
5. Construct a triangle **ABC**, given ∠**ABC** = 62°, ∠**ACB** = 75°, and **AB** − **BC** = 2 cms.; measure **BC**.
6. Inscribe in a given triangle a rectangle having one side double the other.
7. Draw a triangle of sides 5, 6, 7 cms. and construct a square of equal area; measure its side. Check your result from the formula $\sqrt{s(s-a)(s-b)(s-c)}$.
8. Divide a square into three parts of equal area by lines parallel to one diagonal.
9. Given a triangle **ABC**, construct a line parallel to the bisector of ∠**BAC** and bisecting the area of △**ABC**.
10. Given two lines **AB** and **CD**, construct a point **P** on **AB** produced such that **PA** . **PB** = **CD**^{2}.

REVISION PAPERS

BOOK I

I

1. It requires four complete turns of the handle to wind up a bucket from the bottom of a well 24 feet deep. Through what angle must the handle be turned to raise the bucket 5 feet.
2. The angles of a triangle are in the ratio 1 : 3 : 5. Find them.
3. **ACB** is a straight line; **ABX, ACY** are equilateral triangles on opposite sides of **AB**; prove **CX = BY**.
4. **ABCD** is a quadrilateral; **ADCX, BCDY** are parallelograms; prove that **XY** bisects **AB**.

II

5. If the reflex angle **AOB** is four times the acute angle **AOB**, find $\angle$ **AOB**.
6. In $\triangle$ **ABC**, $\angle$ **BAC** = 44°, $\angle$ **ABC** = 112°; find the angle between the lines which bisect $\angle$ **ABC** and $\angle$ **ACB**.
7. The base **BC** of an isosceles triangle **ABC** is produced to **D** so that **CD = CA**, prove $\angle$ **ABD** = 2 $\angle$ **ADB**.
8. **ABCD** is a parallelogram; **P** is the mid-point of **AB**; **CP** and **DA** are produced to meet at **Q**; **DP** and **CB** are produced to meet at **R**; prove **QR = CD**.

III

9. $\angle$ **AOB** = $x°$; **AO** is produced to **C**; **OP** bisects $\angle$ **BOC**; **OQ** bisects $\angle$ **AOB**; calculate reflex angle **POQ**.
10. In $\triangle$ **ABC**, $\angle$ **ABC** = 35°, $\angle$ **ACB** = 75°; the perpendiculars from **B, C** to **AC, AB** cut at **O**. Find $\angle$ **BOC**.

11. The bisector of the angle **BAC** cuts **BC** at **D**; through **C** a line is drawn parallel to **DA** to meet **BA** produced at **P**; prove **AP = AC**.
12. **ABC** is an acute-angled triangle; **BAHK, CAXY** are squares outside the triangle; prove that the acute angle between **BH** and **CX** equals $90° - \angle$ **BAC**.

IV

13. Find the sum of the interior angles of a 15-sided convex polygon.
14. The sum of one pair of angles of a triangle is 100°, and the difference of another pair is 60°; prove that the triangle is isosceles.
15. **ABC** is a triangle right-angled at **C**; **P** is a point on **AB** such that $\angle$ **PCB** = $\angle$ **PBC**; prove $\angle$ **PCA** = $\frac{1}{2}\angle$ **BPC**.
16. **O** is a point inside an equilateral triangle **ABC**; **OAP** is an equilateral triangle such that **O** and **P** are on opposite sides of **AB**; prove **BP = OC**.

V

17. If a ship travels due east or west one sea mile, her longitude alters 1 minute if on the equator, and 2 minutes if in latitude 60°. Find her longitude if she starts (i) at lat. 0°, long. 2° E. and steams 200 miles west; (ii) at lat. 60° N., long. 2° W. and steams 150 miles east.
18. The bisectors of $\angle$s **ABC, ACB** of $\triangle$ **ABC** meet at **O**; if $\angle$ **BOC** = 135°, prove $\angle$ **BAC** = 90°.
19. In $\triangle$ **ABC**, $\angle$ **ACB** = 3 $\angle$ **ABC**; from **AB** a part **AD** is cut off equal to **AC**; prove **CD = DB**.
20. In $\triangle$ **ABC**, **AB = AC**; from any point **P** on **AB** a line is drawn perpendicular to **BC** and meets **CA** produced in **Q**; prove **AP = AQ**.

VI

21. **O** is a point outside a line **ABCD** such that **OA = AB, OB = BC, OC = CD**; $\angle$ **BOC** = $x°$; calculate $\angle$ **OAD** and $\angle$ **ODA** in terms of x.

22. In Fig. 144, page 137, if OQ bisects ∠AOC, prove ∠BOC − ∠BOA = 2∠QOB.
23. ABCD is a quadrilateral; DA = DB = DC; prove ∠BAC + ∠BCA = ½∠ADC.
24. ABCD is a parallelogram; BP, DQ are two parallel lines cutting AC at P, Q; prove BQ is parallel to DP.

VII

25. In △ABC, ∠BAC = 115°, ∠BCA = 20°; AD is the perpendicular from A to BC; prove AD = DB.
26. In Fig. 212, AB is parallel to ED; prove that reflex ∠EDC − reflex ∠ABC = ∠BCD.

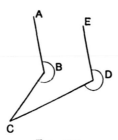

Fig. 212.

27. ABCD is a quadrilateral; ∠ABC = ∠ADC = 90°; prove that the bisectors of ∠s DAB, DCB are parallel.
28. In △ABC, ∠ABC = 90°, ∠ACB = 60°; prove AC = 2BC.

VIII

29. Two equilateral triangles ABC, AYZ lie outside each other; if ∠CAY = 15°, find the angle at which YZ cuts BC.
30. In △ABC, AB = AC; D is a point on AC such that BD = BC; prove ∠DBC = ∠BAC.
31. The altitudes BD, CE of △ABC meet at H; if HB = HC, prove AB = AC.

32. P, Q, R, S are points on the sides AB, BC, CD, DA of a square; if PR is perpendicular to QS, prove PR = QS.

IX

33. In Fig. 213, express x in terms of a, b, c.

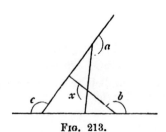

Fig. 213.

34. D is any point on the bisector of ∠BAC; DP, DQ are drawn parallel to AB, AC to meet AC, AB at P, Q; prove DP = DQ.
35. ABC is a △; D, E are points on BC such that ∠BAD = ∠CAE; if AD = AE, prove AB = AC.
36. ABCD is a square; the bisector of ∠BCA cuts AB at P; PQ is the perpendicular from P to AC; prove AQ = PB.

X*

37. ABCDEFGH is a regular octagon; calculate the angle at which AD cuts BF.
38. In △ABC, AD is perpendicular to BC and AP bisects ∠BAC; if ∠ABC > ∠ACB, prove ∠ABC − ∠ACB = 2∠PAD.
39. ABCD is a straight line such that AB = BC = CD; BPQC is a parallelogram; if BP = 2BC, prove PD is perpendicular to AQ.
40. The sides AB, AC of △ABC are produced to D, E; AH, AK are lines parallel to the bisectors of ∠s BCE, CBD meeting BC in H, K: prove AB + AC = BC + HK.

XI*

41. In Fig. 214, express z in terms of a, b, x, y.

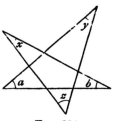

Fig. 214.

42. **AB, BC, CD, DE** are successive sides of a regular n-sided polygon; find the angle between **AB** and **DE**.

43. In △**ABC**, **AB** = **AC**; **BA** is produced to **E**; the bisector of ∠**ACB** meets **AB** at **D**; prove ∠**CDE** = $\frac{3}{4}$ ∠**CAE**.

44. In △**ABC**, ∠**BAC** = 90°; **O** is the centre of the square **BPQC** external to the triangle; prove that **AO** bisects ∠**BAC**.

XII*

45. **B** is 4 miles due east of **A**; a ship sailing from **A** to **B** against the wind takes the zigzag course shown in Fig. 215, her directions being alternately N. 30° E. and S. 30° E.; what is the total distance she travels?

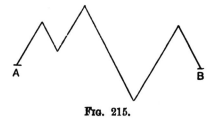

Fig. 215.

46. **ABC** is a triangular sheet of paper, ∠**ABC** = 40°, ∠**ACB** = 75°; the sheet is folded so that **B** coincides with **C**; find the angle which the two parts of **AB** make with each other in the folded position.

47. In △ABC, AB = AC; the bisector of ∠ABC meets AC at D; P is a point on AC produced so that ∠ABP = ∠ADB; prove BC = CP.
48. ABC is a △; BDEC is a square outside △ABC; lines through B, C parallel to AD, AE meet at P; prove PA is perpendicular to BC.

BOOKS I, II

XIII

49. AD, BE are altitudes of △ABC; BC = 5 cms., CA = 6 cms., AD = 4·5 cms.; find BE.
50. ABC is an equilateral triangle; P, Q are points on BC, CA such that BP = CQ; AP cuts BQ at R; prove ∠ARB = 120°.
51. P is a variable point on a circle, centre O, radius a; C is a fixed point at a distance b from O; find the greatest and least possible lengths of CP.
52. ABCD is a quadrilateral; if △ACD = △BCD, prove △ABC = △ABD.

XIV

53. Find in terms of x, y, z the area of Fig. 216.

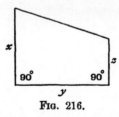

Fig. 216.

54. In △ABC, AB = AC; a line PQR cuts AC produced, AB, BC at R, P, Q; if PQ = QR, prove AP + AR = 2AC.
55. The diagonals of the quadrilateral ABCD cut at O; if △AOD = △BOC, prove △s AOB, COD are equiangular.
56. In △ABC, ∠BAC = 90, AB = 5 cms., AC = 8 cms.; find the area of the triangle and the length of its altitude AD.

XV

57. Find in sq. cms. the area, making any construction and measurements, of Fig. 217.

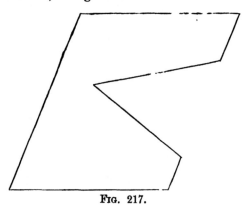

Fig. 217.

58. ABCDE is a regular pentagon; BD cuts CE at P; prove BP = BA.

59. The hypotenuse of a right-angled triangle is $\left(x^2 + \dfrac{1}{x^2}\right)$ inches long, and one of the other sides is $\left(x^2 - \dfrac{1}{x^2}\right)$ inches. Find the third side.

60. The side BC of the parallelogram ABCD is produced to any point K; prove △ABK = quad. ACKD.

XVI

61. ABCD is a parallelogram of area 24 sq. cms.; its diagonals intersect at O; AB = 4·5 cms.; find the distance of O from CD.

62. In △ABC, ∠BAC = 90; BDEC is a square outside △ABC; DX is the perpendicular from D to AC; prove DX = AB + AC.

63. BE, CF are altitudes of △ABC; prove $\dfrac{AB}{AC} = \dfrac{BE}{CF}$.

64. AD is an altitude of △ABC; AB = 7, AC = 5, BC = 8; if BD = x, DC = y, prove $x^2 - y^2 = 24$, and find x, y; find also the area of △ABC.

XVII

65. In Fig. 218, **ABCD** is a quadrilateral inscribed in a rectangle; find the area of **ABCD** in terms of p, q, r, s, x, y.

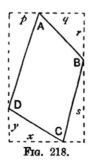

Fig. 218.

66. In $\triangle$**ABC**, $\angle$**BAC** $= 90°$; **P**, **Q** are points on **BC** such that **CA** = **CP** and **BA** = **BQ**; prove $\angle$**PAQ** $= 45°$.
67. **ABCD** is a quadrilateral; $\angle$**ABC** $= \angle$**ADC** $= 90°$; **AP**, **AQ** are drawn parallel to **CD**, **CB**, cutting **CB**, **CD** at **P**, **Q**; prove **QA** . **AB** = **PA** . **AD**. [Use area formulæ.]
68. What is the length of the diagonal of a box whose sides are 3″, 4″, 12″?

XVIII

69. **AD**, **BE**, **CF** are the altitudes of $\triangle$**ABC**; **AB** = $5x$ cms., **BC** = $6x$ cms., **CA** = $3x$ cms., **AD** = 7·5 cms.; find **BE**, **CF**.
70. The base **BC** of the triangle **ABC** is produced to **D**; the lines bisecting $\angle$s **ABD**, **ACD** meet at **P**; a line through **P** parallel to **BC** cuts **AB**, **AC** at **Q**, **R**; prove **QR** = **BQ**~**CR**.
71. **ABCD** is a rhombus; **P**, **Q** are points on **BC**, **CD** such that **BP** = **CQ**; **AP** cuts **BQ** at **O**; prove $\triangle$**AOB** = quad. **OPCQ**.
72. In Fig. 219, **AB** = 2″, **BC** = 4″, **CD** = 1″; if **PD**^{2} = 2**PA**^{2}, find **PB**.

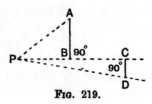

Fig. 219.

XIX

73. Soundings are taken at intervals of 4 feet across a river 40 feet wide, starting 4 feet from one bank, and the following depths in feet are obtained in order 6·6, 9·3, 9·9, 8·2, 8·4, 10·2, 10·5, 7·8, 4·5; find approximately the area of the river's cross-section.

74. In the $\triangle$ABC, AB = BC and $\angle$ABC = 90°; the bisector of $\angle$BAC cuts BC at D; prove AB + BD = AC.

75. ABCD is a parallelogram; P is the mid-point of AD; AB is produced to Q so that AB = BQ; prove ABCD = $2\triangle$PQD.

76. In $\triangle$ABC, $\angle$BAC = 90°; P is the mid-point of AC; PN is drawn perpendicular to BC; prove $BN^2 = BA^2 + CN^2$.

XX

77. ABCD is a parallelogram; AB = $4x$ cms., BC = $5x$ cms.; the distance of A from BC is 6 cms.; find the distance of D from AB.

78. In Fig. 220, AB = BP = 4″, BC = PQ = 3″, AC = BQ = 5″; calculate the area common to $\triangle$s ABC, BPQ.

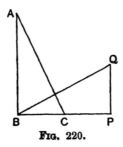

Fig. 220.

79. In $\triangle$ABC, AB = AC; P is any point on BC; Q, R are the mid-points of BP, PC; QX, RY are drawn perpendicular to BC and cut AB, AC at X, Y; prove BX = AY.

80. ABC is an equilateral triangle; BC is bisected at D and produced to E so that CE = CD, prove $AE^2 = 7EC^2$.

XXI

81. In Fig. 221, the triangle **ABC** is inscribed in a rectangle: find its area and the distance of **A** from the mid-point of **BC**.

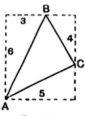

Fig. 221.

82. **A, B** are fixed points; **X** is a variable point such that $\angle$ **AXB** is obtuse; the perpendicular bisectors of **AX, BX** cut **AB** at **Y, Z**; prove that the perimeter of $\triangle$ **XYZ** is constant.

83. **ABC** is a $\triangle$; a line **XY** parallel to **BC** cuts **AB, AC** at **X, Y** and is produced to **Z** so that **XZ = BC**; prove $\triangle$ **BXY** = $\triangle$ **AYZ**.

84. The sides of a triangle are 8 cms., 9 cms., 12 cms. Is it obtuse-angled?

XXII*

85. **ABC** is a triangle of area 24 sq. cms.; **AB** = 8 cms., **AC** = 9 cms.; **D** is a point on **BC** such that **BD** = $\tfrac{1}{3}$**BC**; find the distance of **D** from **AB**.

86. **O** is a point inside $\triangle$**ABC** such that **OA** = **AC**, prove that **BA** > **AC**.

87. **ABCD** is a quadrilateral; **AB** is parallel to **CD**; **BP, CP** are drawn parallel to **AC, AD** to meet at **P**; prove $\triangle$ **PDC** = **ABD**.

88. The length, breadth, and height of a room are each 10 feet; **CAE, DBF** are two vertical lines bisecting opposite walls, **C, D** being on the ceiling and **E, F** on the floor; **CA** = x feet, **DB** = 4 feet. Find in terms of x the shortest path from **A** to **B**—(i) along these two walls and the ceiling; (ii) along these two walls and one other wall. What is the condition that route (ii) is shorter than route (i)?

XXIII

89. In Fig. 222, AB = 9″, BC = 8″, CD = 7″; if AP PD, calculate BP.

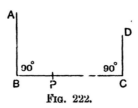
Fig. 222.

90. ABC is a △; AP is the perpendicular from A to the bisector of ∠ABC; PQ is drawn parallel to BC to cut AB at Q; prove AQ = QB = PQ.

91. ABP, ABQ are equivalent triangles on opposite sides of AB; PR is drawn parallel to BQ to meet AB at R; prove QR is parallel to PB.

92. In △ABC, ∠BAC = 90°; H, K are the mid-points of AB, AC; prove that $BH^2 + HK^2 + KC^2 = \tfrac{1}{2}BC^2$.

XXIV*

93. The angles at the corners of Fig. 223 are all right angles. Construct a line parallel to AB to bisect the given figure. [The fact in Ex. XXXI, No. 22, may be useful.]

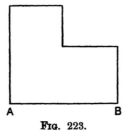
Fig. 223.

94. In △ABC, ∠BAC = 90°; P, Q are the centres of the two squares which can be described on BC; prove that the distances of P, Q from AB are $\tfrac{1}{2}(AB \pm AC)$.

95. ABCD is a parallelogram; any line parallel to BA cuts BC, AC, AD at X, Y, Z; prove $\triangle AXY = \triangle DYZ$.
96. In $\triangle ABC$, $\angle ACB = 90°$; AD is a median; prove that $AB^2 = AD^2 + 3BD^2$.

BOOKS I–III

XXV

97. The side BC of an equilateral triangle ABC is produced to D so that $CD = 3BC$; prove $AD^2 = 13AB^2$.
98. ABCD is a quadrilateral; if $\angle ABC + \angle ADC = 180°$, prove that the perpendicular bisectors of AC, BD, AB are concurrent.
99. ABCD is a quadrilateral inscribed in a circle; AC is a diameter; $\angle BAC = 43°$; find $\angle ADB$.
100. Two circles ABPQ, ABR intersect at A, B; BP is a tangent to circle ABR; RAQ is a straight line; prove PQ is parallel to BR.

XXVI

101. ABC is a $\triangle$; H, K are the mid-points of AB, AC; P, Q are points on BC such that $BP = \tfrac{1}{4}BC = \tfrac{1}{3}BQ$; prove $PH = QK$.
102. Find the remaining angles in Fig. 224.

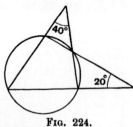

Fig. 224.

103. ABCD is a parallelogram; the circle through A, B, C cuts CD at P; prove $AP = AD$.
104. APB, AQB are two circles; AP is a tangent to circle AQB; PBQ is a straight line; prove that AQ is parallel to the tangent at P.

REVISION PAPERS

XXVII

105. ABCD is a square; P is a point on AB such that $AP = \tfrac{1}{3}AB$; Q is a point on PC such that $PQ = \tfrac{1}{3}PC$; prove $APQD = \tfrac{1}{2}ABCD$.
106. AOB is a diameter of a circle perpendicular to a chord POQ; $AO = h$, $PQ = a$; find AB in terms of a, h.
107. The side AB of a cyclic quadrilateral ABCD is produced to E; $\angle DBE = 140°$, $\angle ADC = 100°$, $\angle ACB = 45°$; find $\angle BAC$, $\angle CAD$.
108. In $\triangle ABC$, $\angle BAC = 90°$; the circle on AB as diameter cuts BC at D; the tangent at D cuts AC at P; prove $PD = PC$.

XXVIII

109. In quadrilateral ABCD, $AB = 7''$, $CD = 11''$, $\angle BAD = \angle ADC = 90°$, $\angle BCD = 60°$; calculate AC.
110. Two chords AB, DC of a circle, centre O, are produced to meet at E; $\angle CBE = 75°$, $\angle CEB = 22°$, $\angle AOD = 144°$; prove $\angle AOB = \angle BAC$.
111. In Fig. 225, O is the centre and TQ bisects $\angle OTP$; prove $\angle TQP = 45°$.

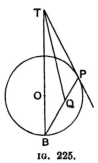

Fig. 225.

112. PAB, PBC, PCA are three unequal circles; from any point D on the circle PBC, lines DB, DC are drawn and produced to meet the circles PBA, PCA again at X, Y; prove XAY is a straight line.

XXIX

113. In $\triangle ABC$, $\angle ACB = 90°$, $AC = 2CB$; CD is an altitude; prove by using the figure of Pythagoras' theorem or otherwise that $AD = 4DB$.

114. In Fig. 226, O is the centre of the circle; PQ and PT are equally inclined to TO; prove $\angle QOT = 3\angle POT$.

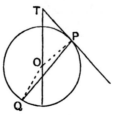

Fig. 226.

115. AOB is a chord of a circle ABC; T is a point on the tangent at A; the tangent at B meets TO produced at P; $\angle ATO = 35°$, $\angle BOT = 115°$; find $\angle BPT$.

116. In $\triangle ABC$, $AB = AC$; the circle on AB as diameter cuts BC at P; prove $BP = PC$.

XXX

117. X, Y, Z are any points on the sides BC, CA, AB of the triangle ABC; prove that $AX + BY + CZ > \frac{1}{2}(BC + CA + AB)$.

118. A, B, C, D are the first milestones on four straight roads running from a town X; A is due north of D and north-west of B. C is E. 20° S. of D; find the bearing of B from C.

119. $ABCD$ is a quadrilateral inscribed in a circle, centre O; if AC bisects $\angle BAD$, prove that OC is perpendicular to BD.

120. A diameter AB of a circle APB is produced to any point T; TP is a tangent; prove $\angle BTP + 2\angle BPT = 90°$.

XXXI

121. $ABCD$ is a rectangle; P is any point on CD; prove that quad. $ABCP - \triangle APD = AD \cdot CP$.

122. $ABCD$ is a circle; if arc $ABC = \frac{1}{4}$ arc ADC, find $\angle ADC$.

REVISION PAPERS

123. A, B, C are points on a circle, centre O; BO, CO are produced to meet AC, AB at P, Q; prove $\angle BPC + \angle BQC = 3 \angle BAC$.

124. In Fig. 227, AB is a diameter; $\angle HPQ = \angle KQF = 90°$; prove AH = BK.

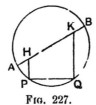

Fig. 227.

XXXII

125. In $\triangle ABC$, $\angle BAC = 90°$; AD is an altitude; prove that
$$\frac{1}{AD^2} = \frac{1}{AB^2} + \frac{1}{AC^2}.$$

126. ABCD is a square inscribed in a circle; P is any point on the minor arc AB; prove $\angle APB = 3 \angle BPC$.

127. ABC is a triangle inscribed in a circle; the bisector of $\angle BAC$ meets the circle at P; I is a point on PA between P and A such that PI = PB; prove $\angle IBA = \angle IBC$.

128. Two circles, centres A, B, cut at X, Y; XP, XQ are the tangents at X; prove $\angle AXB$ is equal or supplementary to $\angle PXQ$.

XXXIII

129. ABCD is a parallelogram; P is any point on CD; PA, PB, CB, AD cut any line parallel to AB at X, Y, Z, W; prove $DCZW = 2 \triangle APY$.

130. In Fig. 228, O is the centre, PQ = AO, $\angle AOQ = 90°$; prove arc BR = 3 arc AP.

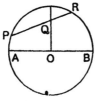

Fig. 228.

131. A rectangular strip of cardboard is 7 inches wide, 4 feet long; how many circular discs each of radius 2 inches can be cut out of it ?

132. AB, CD are parallel chords of a circle ABDC, centre O; prove $\angle$ AOC equals angle between AD and BC.

XXXIV

133. Two metre rules AOB, COD cross one another at right angles: the zero graduations are at A, C; a straight edge XY, half a metre long, moves with one end X on OB and the other end Y on OD; when the readings for X are 50, 40 cms., those for Y are 50, 60 cms. respectively. Find the readings at O.

134. Two circles PARB, QASB intersect at A, B; a line PQRS cuts one at P, R and the other at Q, S; prove $\angle$ PAQ = $\angle$ RBS.

135. In $\triangle$ ABC, $\angle$ BAC = 90°; D is the mid-point of BC; a circle touches BC at D, passes through A and cuts AC again at E; prove arc AD = 2 arc DE.

136. Two circular cylinders of radii 2″, 6″ are bound tightly together with their axes parallel by an elastic band. Find its stretched length.

XXXV

137. In Fig. 229, BC is an arc of radius 8″ whose centre lies on OB produced; OB = 9″, $\angle$ AOB = 90°; calculate the radius of a circle touching AO, OB and arc BC.

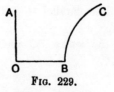

Fig. 229.

138. ABCD is a parallelogram; AB, CB are produced to X, Y; P is any point within the angle XBY; prove $\triangle$ PCD − $\triangle$ PAB = $\triangle$ ABC.

139. $A_1 A_2 A_3 \ldots A_{20}$ is a reguar polygon of 20 sides, prove that $A_1 A_8$ is perpendicular to $A_3 A_{16}$.

140. A, B, C are three points on a circle; the tangent at A meets BC produced at D; prove that the bisectors of ∠s BAC, BDA are at right angles.

XXXVI

141. In △ABC, ∠ABC = 90°, ∠BAC = 1.°, the bisector of ∠ACB meets AB at P; prove $AP^2 = 2PB^2$.
142. The diameter AB of a circle is produced to any point P; a line is drawn from P touching the circle at Q and cutting the tangent at A in R; prove ∠BQP = ½∠ARP.
143. In △ABC, AB = AC and ∠BAC is obtuse; a circle is drawn touching AC at A, passing through B and cutting BC again at P; prove arc AB = 2 arc AP.
144. The volume of a circular cylinder is V cub. in. and the area of its curved surface is S sq. in.; find its radius in terms of V, S.

BOOKS I–IV

XXXVII

145. In Fig. 230, if ∠ADC = ∠BEA = ∠CFB, prove that the triangles ABC, XYZ are equiangular.

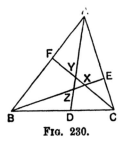

Fig. 230.

146. The tangent at a point R of a circle meets a chord PQ at T; O is the centre; E is the mid-point of PQ; prove ∠ROT = RET.
147. A line AB, 8 cms. long, is divided internally and externally in the ratio 3 : 1 at P, Q respectively; find PQ : AB.

148. ABCD is a quadrilateral; a line AF parallel to BC meets BD at F; a line BE parallel to AD meets AC at E; prove EF is parallel to CD.

XXXVIII

149. The sides AB, BC, CA of △ABC are produced their own lengths to X, Y, Z; prove △XYZ = 7△ABC.
150. ABCD is a quadrilateral; the circles on AB, BC as diameters intersect again at P; the circles on AD, DC as diameters intersect again at Q; prove BP is parallel to DQ.
151. A town occupies an oval area of length 2400 yards, breadth 1000 yards: a plan is made of it on a rectangular sheet of paper 18" long, 12" wide. What is the best scale to choose?
152. ABC is a triangle inscribed in a circle; AD is an altitude; AP is a diameter; prove $\dfrac{AB}{AP} = \dfrac{AD}{AC}$ and complete the equation $\dfrac{BD}{AB} = \dfrac{}{AP}$.

XXXIX

153. AB is a diameter of a circle; AOC, BOE are two chords such that ∠CAB = ∠EBA = $22\tfrac{1}{2}°$; prove that $AO^2 = 2OC^2$.
154. PQ is a chord of a circle; T is a point on the tangent at P such that PT = PQ; TQ cuts the circle at R; prove ∠RPT = $60° ± \tfrac{1}{3}$ ∠QPR.
155. In Fig. 231, AB, CD, EF are parallel; AD = 7", DF = 3", CE = 4"; find BC. If EF = 2", AB = 3", find CD.

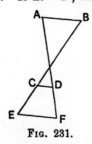

Fig. 231.

156. AB, DC are parallel sides of the trapezium ABCD; AC cuts DB at O; the line through O parallel to AB cuts AD, BC at P, Q; prove PO = OQ.

XL

157. In $\triangle ABC$, $AB = AC$ and $\angle BAC = 120°$; the perpendicular bisector of AB cuts BC at X; prove $BC = 3BX$.
158. AOB, COD are two perpendicular chords of a circle; prove that arc AC + arc BD equals half the circumference.
159. A light is placed 4′ in front of a circular hole 3″ in diameter in a partition; find the diameter of the illuminated part of a wall 5′ behind the partition and parallel to it.
160. ABC is a triangle inscribed in a circle; $AB = AC$; AP is a chord cutting BC at Q; prove $AP \cdot AQ = AB^2$.

XLI

161. In $\triangle ABC$, $\angle BAC = 90°$, $\angle ABC = 45°$; AB is produced to D so that $AD \cdot DB = AB^2$; prove that the perpendicular bisector of CD bisects AB.
162. $ABCD$ is a cyclic quadrilateral; AC cuts BD at O; if CD touches the circle OAD, prove that CB touches the circle OAB.
163. $ABCDEF$ is a straight line; $AB : BC : CD : DE : EF = 2 : 3 : 7 : 4 : 5$; find the ratios $\dfrac{AD}{DF}$ and $\dfrac{BE}{AF}$.
164. $ABCD$ is a parallelogram; a line through A cuts BD, BC, CD at E, F, G; prove $\dfrac{AE}{EF} = \dfrac{AG}{AF}$.

XLII

165. AB is a diameter of a circle APB; the tangent at A meets BP at Q; prove that the tangent at P bisects AQ.
166. PAQ, PBQ, PCQ are three equal angles on the same side of PQ; the bisectors of $\angle$s PAQ, PBQ meet at H; prove that CH bisects $\angle PCQ$.
167. Two triangles are equiangular: the sides of one are 3 cms., 5 cms., 7 cms.; the perimeter of the other is $2\frac{1}{2}$ feet; find its sides.
168. Two lines OAB, OCD cut a circle at A, B, C, D; H, K are points on OB, OD such that $OH = OC$, $OK = OA$; prove that HK is parallel to BD.

XLIII

169. C is the mid-point of AB; P is any point on CB; prove that $AP^2 - PB^2 = 2AB \cdot CP$.
170. A circular cylinder of height 6" is cut from a sphere of radius 4"; find its greatest volume.
171. Show that the triangle whose vertices are (2, 1), (5, 1), (4, 2) is similar to the triangle whose vertices are (1, 1) (7, 1) (5, 3).
172. Two circles intersect at A, B; the tangents at A meet the circles at C, D; prove $\dfrac{BC}{BA} = \dfrac{BA}{BD}$.

XLIV

173. ABCD is a quadrilateral; AP is drawn equal and parallel to BD; prove $\triangle APC$ = quad. ABCD.
174. A circular cone is made from a sector of a circle of radius 6" and angle 240°; find its height.
175. A straight rod AB, 3' 9" long, is fixed under water with A 2' 6" and B 9" below the surface; what is the depth of a point C on the rod where AC = 1'?
176. ABCD is a straight line; O is a point outside it; a line through B parallel to OD cuts OA, OC at P, Q; if PB = BQ, prove $\dfrac{AB}{BC} = \dfrac{AD}{CD}$.

XLV

177. In Fig. 232, OA, AB are two rods hinged together at A; the end O is fixed, and AO can turn freely about it; the end B is constrained to slide in a fixed groove OC.

OA = 3', AB = 4'; find the greatest length of the groove which B can travel over, and calculate the distance of B from O when AB makes the largest possible angle with OC.

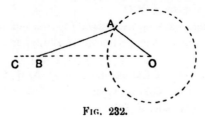

Fig. 232.

178. ABC is a triangle inscribed in a circle; P, Q, R are the midpoints of the arcs BC, CA, AB; prove AP is perpendicular to QR.
179. AOXB, COYD are two straight lines; AC, XY, BD are parallel lines cutting them; AX = 7, XB = 3, AC = 2, BD = 4; find XY.
180. P is any point on the common chord of two circles, centres A, B; HPK and XPY are chords of the two circles perpendicular to PA, PB respectively; prove HK = XY.

XLVI

181. ABC is a triangle inscribed in a circle; the internal and external bisectors of $\angle$ BAC cut BC at P, Q; prove that the tangent at A bisects PQ.
182. A circle of radius 4 cms. touches two perpendicular lines; calculate the radius of the circle touching this circle and the two lines.
183. ABCD is a rectangle; AB = 8″, BC = 5″; P is a point inside it whose distances from AD, AB are 2″, 1″; DP is produced to meet AB at E; CE cuts AD at F; calculate EB, AF.
184. Two lines OAB, OCD meet a circle at A, B, C, D; prove that $\dfrac{OA \cdot OD}{OB \cdot OC} = \dfrac{AD^2}{BC^2}$.

XLVII*

185. ABC is an equilateral triangle; P is any point on BC; AC is produced to Q so that CQ = BP; prove AP = PQ.
186. AB is a diameter of a circle APB; AH, BK are the perpendiculars from A, B to the tangent at P; prove that AH + BK = AB.
187. A chord AB of a circle ABT is produced to O; OT is a tangent; OA = 6″, OT = 4″, AT = 3″, find BT.
188. AB, DC are parallel sides of the trapezium ABCD; AC cuts BD at E; DA, CB are produced to meet at F; EF cuts AB, DC at P, Q; prove $\dfrac{QE}{EP} = \dfrac{QF}{PF}$.

XLVIII*

189. A brick rests on the ground and an equal brick is propped up against it as in Fig. 233. The bricks are 4″ by 2″. Calculate the height of each corner of the second brick above the ground, if AB = $1\frac{1}{2}$″.

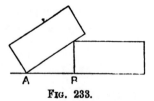

Fig. 233.

190. Prove that the area of a square inscribed in a given semicircle is $\frac{2}{5}$ of the area of the square inscribed in the whole circle.
191. The bisector of ∠ BAC cuts BC at D; the line through D perpendicular to DA cuts AB, AC at Y, Z; prove $\frac{BY}{CZ} = \frac{BD}{DC}$.
192. A chord AD is parallel to a diameter BC of a circle; the tangent at C meets AD at E; prove BC . AE = BD².

XLIX*

193. A is a fixed point on a given circle; a variable chord AP is produced to Q so that PQ is of constant length; QR is drawn perpendicular to AQ; prove that QR touches a fixed circle.
194. Four equal circular cylinders, diameter 4″, length 5″, are packed in a rectangular box; what is the least amount of unoccupied space in the box?
195. A rectangular sheet of paper ABCD is folded so that B falls on CD and the crease passes through A; AB = 10″, BC = 6″; find the distance of the new position of B from C. If the crease meets BC at Q, find CQ.
196. ABCD is a parallelogram; a line through A cuts BD, CD, BC in P, Q, R; prove $\frac{PQ}{PR} = \frac{PD^2}{PB^2}$.

L

197. In Fig. 234, **ABCD** is a rectangle; $BP = 2CQ$; $AD = 2AB = 6''$. The area of **APQD** is 10 sq. in.; find **BP**.

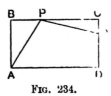

Fig. 234.

198. **ABC** is a triangle inscribed in a circle; the tangents at **B, C** meet at **T**; a line through **T** parallel to the tangent at **A** meets **AB, AC** produced at **D, E**; prove $DT = TE$.

199. A line **HK** parallel to **BC** cuts **AB, AC** at **H, K**; the distance between **HK** and **BC** is 5 cms.; the areas of **AHK** and **HKCB** are 9 sq. cms., 40 sq. cms.; find **HK**.

200. In $\triangle ABC$, **I** is the in-centre and I_1 is the ex-centre corresponding to **BC**; prove $AI \cdot AI_1 = AB \cdot AC$.

WHEN learning propositions, do not use the figure printed in the book, but **draw your own figure** instead.

It is more trouble but gives better results. For this reason, no attempt has been made to arrange the whole proof of every theorem on the same page as the corresponding figure.

A freehand figure is good enough.

PROOFS OF THEOREMS

BOOK I

DEFINITION.—If **C** is any point on the straight line **AB**, and if a line **CD** is drawn so that the angles **ACD**, **BCD** are equal, each is called a *right angle*.

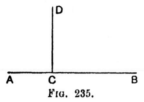
Fig. 235.

Therefore if **C** is any point on the straight line **AB**, the angle **ACB** is equal to two right angles, or 180°.

Theorem 1

(1) If one straight line stands on another straight line, the sum of the two adjacent angles is two right angles.
(2) If at a point in a straight line, two other straight lines, on opposite sides of it, make the adjacent angles together equal to two right angles, these two straight lines are in the same straight line.

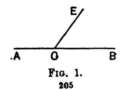
Fig. 1.

(1) *Given* **CE** meets **AB** at **C**.
  To Prove ∠ **ACE** + ∠ **BCE** = 180°.
  ∠ **ACE** + ∠ **BCE** = ∠ **ACB**
        = 180°, since **ACB** is a st. line.
               Q.E.D.

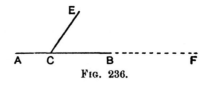

FIG. 236.

(2) *Given* ∠ **ACE** + ∠ **BCE** = 180°.
  To Prove **ACB** is a straight line.
  Produce **AC** to **F**.
  ∴ ∠ **ACE** + ∠ **FCE** = 180°, since **ACF** is a st. line.
  But ∠ **ACE** + ∠ **BCE** = 180°, given.
  ∴   ∠ **ACE** + ∠ **FCE** = ∠ **ACE** + ∠ **BCE**.
    ∴   ∠ **FCE** = ∠ **BCE**.
    ∴   **CB** falls along **CF**.
  But **ACF** is a st. line; ∴ **ACB** is a st. line.
               Q.E.D.

THEOREM 2

If two straight lines intersect, the vertically opposite angles a equal.

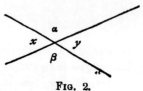

FIG. 2.

To Prove that $x = y$ and $a = \beta$.
    $x + a = 180°$ adjacent angles.
    $a + y = 180°$ adjacent angles.
  ∴ $x + a = a + y$.
  ∴ $x = y$.
  Similarly $a = \beta$.
               Q.E.D.

For riders on Theorems 1–2, see page 2.

PROOFS OF THEOREMS

THEOREM 3

If two triangles have two sides of one equal respectively to two sides of the other, and if the included angles are equal, then the triangles are congruent.

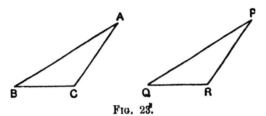

Fig. 23.

Given AB = PQ, AC = PR, ∠ BAC = ∠ QPR.
To Prove △ ABC ≡ △ PQR.
 Apply the triangle ABC to the triangle PQR, so that A falls on P and the line AB along the line PQ;
 Since AB = PQ, ∴ B falls on Q.
 Also since AB falls along PQ and ∠ BAC = ∠ QPR, ∴ AC falls along PR.
 But AC = PR, ∴ C falls on R.
 ∴ the triangle ABC coincides with the triangle PQR.
 ∴ △ABC ≡ △PQR.

Q.E.D.

For riders on Theorems 3, 9, 10, see page 16.

THEOREM 4

If one side of a triangle is produced, the exterior angle is greater than either of the interior opposite angles.

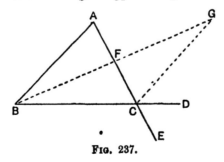

Fig. 237.

BC is produced to **D**.

To Prove ∠ **ACD** > ∠ **ABC** and ∠ **ACD** > ∠ **BAC**.

Let **F** be the middle point of **AC**. Join **BF** and produce it to **G**, so that **BF** = **FG**. Join **CG**.

In the triangles **AFB, CFG**

AF = **FC** and **BF** = **FG**, constr.

∠ **AFB** = ∠ **CFG**, vert. opp.

∴ △**AFB** ≡ △**CFG**.

∴ ∠ **BAF** = ∠ **GCF**.

But ∠ **DCA** > its part ∠ **GCF**.

∴ ∠ **DCA** > ∠ **BAF** or ∠ **BAC**.

Similarly, if **BC** is bisected and if **AC** is produced to **E**, it can be proved that ∠ **BCE** > ∠ **ABC**.

But ∠ **ACD** = ∠ **BCE**, vert. opp.

∴ ∠ **ACD** > ∠ **ABC**.

Q.E.D.

DEFINITION.—Straight lines which *lie in the same plane* and which never meet, however far they are produced either way, are called *parallel* straight lines.

PLAYFAIR'S AXIOM.—Through a given point, one and only one straight line can be drawn parallel to a given straight line.

THEOREM 5

If one straight line cuts two other straight lines such that
  either (1) the alternate angles are equal,
  or   (2) the corresponding angles are equal,
  or   (3) the interior angles on the same side of the cutting line are supplementary,
then the two straight lines are parallel.

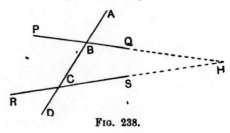

FIG. 238.

PROOFS OF THEOREMS 209

ABCD cuts PQ, RS at B, C.
(1) *Given* ∠ PBC = ∠ BCS.
  To Prove PQ is parallel to RS.
  If PQ, RS are not parallel, they will meet when produced, at H, say.
  Since BCH is a triangle,
    ext. ∠ PBC > int. ∠ BCH,
  which is contrary to hypothesis.
    ∴ PQ cannot meet RS and is ∴ parallel to it.
                                           Q.E.D.
(2) *Given* ∠ ABQ = ∠ BCS.
  To Prove PQ is parallel to RS.
    ∠ ABQ = ∠ PBC, vert. opp.
   But ∠ ABQ = ∠ BCS, given.
    ∴ ∠ PBC = ∠ BCS.
    ∴ by (1), PQ is parallel to RS.
(3) *Given* ∠ QBC + ∠ SCB = 180°.
  To Prove PQ is parallel to RS.
    ∠ QBC + ∠ PBC = 180°, adj. angles.
   But ∠ QBC + ∠ SCB = 180°, given.
    ∴ ∠ QBC + ∠ PBC = ∠ QBC + ∠ SCB.
    ∴ ∠ PBC = ∠ SCB.
    ∴ by (1), PQ is parallel to RS.
                                           Q.E.D.

THEOREM 6

If a straight line cuts two parallel straight lines,
  Then (1) the alternate angles are equal;
       (2) the corresponding angles are equal;
       (3) the interior angles on the same side of the cutting line are supplementary.

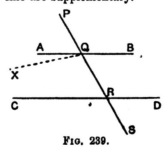

Fig. 239.

AB, CD are two parallel st. lines; the line **PS** cuts them at **Q, R**.

To Prove (1) ∠ **AQR** = ∠ **QRD**.
            (2) ∠ **PQB** = ∠ **QRD**.
            (3) ∠ **BQR** + ∠ **QRD** = 180°.

(1) If ∠ **AQR** is not equal to ∠ **QRD**, let the angle **XQR** be equal to ∠ **QRD**.

But these are alternate angles.

∴ **QX** is parallel to **RD**,

∴ two intersecting lines **QX, QA** are both parallel to **RD**, which is impossible by Playfair's Axiom.

∴ ∠ **AQR** cannot be unequal to ∠ **QRD**.
∴ ∠ **AQR** = ∠ **QRD**.

(2)   ∠ **PQB** = ∠ **AQR**, vert. opp.
But ∠ **AQR** = ∠ **QRD**, alt. angles.
∴ ∠ **PQB** = ∠ **QRD**.

(3)   ∠ **BQR** + ∠ **AQR** = 180°, adj. angles.
But   ∠ **AQR** = ∠ **QRD**, alt. angles.
∴ ∠ **BQR** + ∠ **QRD** = 180°.

Q.E.D.

For riders on Theorems 5, 6, see page 6.

Theorem 7

(1) If a side of a triangle is produced, the exterior angle is equal to the sum of the two interior opposite angles.
(2) The sum of the three angles of any triangle is two right angles.

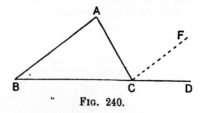

Fig. 240.

ABC is a triangle; **BC** is produced to **D**.

To Prove (1) ∠ **ACD** = ∠ **CAB** + ∠ **ABC**.
            (2) ∠ **CAB** + ∠ **ABC** + ∠ **ACB** = 180°.

(1) Let **CF** be drawn parallel to **AB**.

$$\angle FCD = \angle ABC, \text{ corresp. angles.}$$
$$\angle ACF = \angle CAB, \text{ alt. angles.}$$
adding, $\angle FCD + \angle ACF = \angle ABC + \angle CAB.$
$$\therefore \angle ACD = \angle ABC + \angle CAB.$$

(2) Add to each the angle **ACB**.

$$\therefore \angle ACD + \angle ACB = \angle ABC + \angle CAB + \angle ACB.$$
But $\angle ACD + \angle ACB = 180°,$ adj. angles.
$$\therefore \angle ABC + \angle CAB + \angle ACB = 180°.$$

Q.E.D.

Theorem 8

(1) All the interior angles of a convex polygon, together with four right angles, are equal to twice as many right angles as the polygon has sides.
(2) If all the sides of a convex polygon are produced in order, the sum of the exterior angles is four right angles.

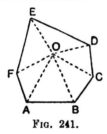

Fig. 241.

Let n be the number of sides of the polygon.

(1) *To Prove* that

the sum of the angles of the polygon $+ 4$ rt. $\angle$ s $= 2n$ rt. $\angle$ s.

Take any point **O** inside the polygon and join it to each vertex.

The polygon is now divided into n triangles.

But the sum of the angles of each triangle is 2 rt. $\angle$ s.

$\therefore$ the sum of the angles of the n triangles is $2n$ rt. $\angle$ s.

But these angles make up all the angles of the polygon together with all the angles at **O**.

212   CONCISE GEOMETRY

Now the sum of all the angles at O is 4 rt. ∠ s.
∴ all the angles of the polygon + 4 rt. ∠ s = 2n rt. ∠ s.

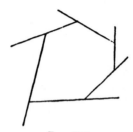

Fig. 242.

(2) At each vertex, the interior ∠ + the exterior ∠ = 2 rt. ∠ s.
∴ the sum of all the interior angles + the sum of all the exterior angles = 2n rt. ∠ s.
But the sum of all the interior angles + 4 rt. ∠ s = 2n rt. ∠ s.
∴ the sum of all the exterior ∠ s = 4 rt. ∠ s.

Q.E.D.

Theorem 8(1) may also be stated as follows :—
The sum of the interior angles of any convex polygon of n sides is $2n - 4$ right angles.

For riders on Theorems 7, 8, see page 10.

Theorem 9

Two triangles are congruent if two angles and a side of one are respectively equal to two angles and the corresponding side of the other.

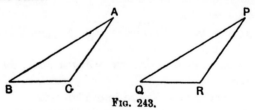

Fig. 243.

Given either that BC = QR.
∠ ABC = ∠ PQR.
∠ ACB = ∠ PRQ.

PROOFS OF THEOREMS 213

or that BC = QR.
∠ABC = ∠PQR.
∠BAC = ∠QPR.
To Prove △ABC ≡ △PQR.
The sum of the three angles of any triangle is 180°.
∴ in each case, the remaining pair of angles is equal.
Apply the triangle ABC to the triangle PQR so that B falls on Q and BC falls along QR.
Since BC = QR, C falls on R.
And since BC falls on QR and ∠ABC = ∠PQR, ∴ BA falls along QP.
And since CB falls on RQ and ∠ACB = ∠PRQ, ∴ CA falls along RP.
∴ A falls on P.
∴ the triangle ABC coincides with the triangle PQR.
∴ △ABC ≡ △PQR.

Q.E.D.

THEOREM 10

(1) If two sides of a triangle are equal, then the angles opposite to those sides are equal.
(2) If two angles of a triangle are equal, then the sides opposite to those angles are equal.

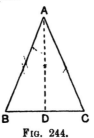

FIG. 244.

ABC is a triangle : let the line bisecting the angle BAC meet BC at D.

(1) *Given* AB = AC.
To Prove ∠ACB = ∠ABC.
In the △s ABD, ACD.
AB = AC, given.
AD is common.

∠ BAD = ∠ CAD, constr.
∴ the △s are congruent.
∴ ∠ ABD = ∠ ACD.
(2) *Given* ∠ ABC = ∠ ACB.
To Prove AC = AB.
In the △s ABD, ACD.
∠ ABD = ∠ ACD, given.
∠ BAD = ∠ CAD, constr.
AD is common.
∴ the △s are congruent.
∴ AB = AC.

Q.E.D.

For riders on Theorems 3, 9, 10 see page 15.

Theorem 11

Two triangles are congruent if the three sides of one are respectively equal to the three sides of the other.

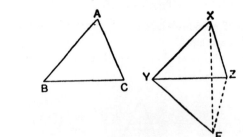

Fig. 245(1).

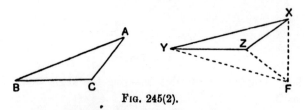

Fig. 245(2).

Given that AB = XY, BC = YZ, CA = ZX.
To Prove △ ABC ≡ △ XYZ.
Place the triangle ABC so that B falls on Y and BC along YZ; ∴ since BC = YZ, C falls on Z.

PROOFS OF THEOREMS 215

Let the point A fall at a point F on the opposite side of YZ to X. Join XF.
Now YF = BA, constr.
But BA = YX, given.
∴ YF = YX
But these are sides of the triangle YFX.
∴ ∠YXF = ∠YFX.
Similarly, ∠ZXF = ∠ZFX.
∴ adding in Fig. 245(1) or subtracting in Fig. 245(2)
∠YXZ = ∠YFZ.
But ∠BAC = ∠YFZ, constr.
∴ ∠BAC = ∠YXZ.
∴ in the △s ABC, XYZ
AB = XY, given.
AC = XZ, given.
∠BAC = ∠YXZ, proved.
∴ △ABC ≡ △XYZ.

Q.E.D.

THEOREM 12

Two right-angled triangles are congruent if the hypotenuse and side of one are respectively equal to the hypotenuse and a side of the other.

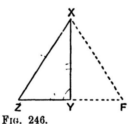

FIG. 246.

Given ∠ABC = 90° = ∠XYZ.
AC = XZ.
AB = XY.
To Prove △ABC ≡ △XYZ.

Place the triangle ABC so that A falls on X and AB falls along XY, and so that C falls at some point F on the opposite side of XY to Z.

Since **AB = XY**, **B** falls on **Y**.
∠ **XYF** = ∠ **ABC** = 90° and ∠ **XYZ** = 90°.
∴   ∠ **XYF** + ∠ **XYZ** = 180°.
∴   **ZYF** is a straight line.
But **XF** = **AC**, and **AC** is given equal to **XZ**.
∴   **XZF** is a triangle, in which **XF** = **XZ**.
∴   ∠ **XZY** = ∠ **XFY**.
But ∠ **XFY** = ∠ **ACB**, constr.
∴   ∠ **XZY** = ∠ **ACB**.
∴   in the △s **XYZ, ABC**.
∠ **XYZ** = ∠ **ABC**, given.
∠ **XZY** = ∠ **ACB**, proved.
XY = **AB**, given.
∴   △ **XYZ** ≡ △ **ABC**.

Q.E.D.

Theorem 13

(1) The opposite sides and angles of a parallelogram are equal.
(2) Each diagonal bisects the parallelogram.

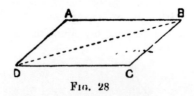

Fig. 28

Given **ABCD** is a parallelogram.
To Prove (1)   **AB** = **CD** and **AD** = **BC**.
∠ **DAB** = ∠ **DCB** and ∠ **ABC** = ∠ **ADC**.
(2) **AC** and **BD** each bisect the parallelogram.
Join **BD**.
In the △s **ADB, CBD**
∠ **ADB** = ∠ **CBD**, alt. ∠ s.
∠ **ABD** = ∠ **CDB**, alt. ∠ s.
BD is common.
∴   △ **ADB** ≡ △ **CBD**.
∴   **AB** = **CD**, **AD** = **BC**, ∠ **DAB** = ∠ **BCD**
and **BD** bisects the parallelogram.

PROOFS OF THEOREMS 217

Similarly, by joining **AC** it may be proved that ∠ **ABC** = ∠ **ADC**, and that **AC** bisects the parallelogram.

Q.E.D.

Theorem 14

The diagonals of a parallelogram bisect one another.

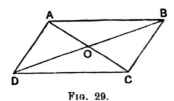

Fig. 29.

The diagonals **AC, BD** of the parallelogram **ABCD** intersect at **O**.
To Prove **AO = OC** and **BO = OD**.
In the △s **AOD, COB**,
   ∠ **DAO** = ∠ **BCO**, alt. ∠ s.
   ∠ **ADO** = ∠ **CBO**, alt. ∠ s.
   AD = BC, opp. sides of ||gram.
∴ △ **AOD** ≡ △ **COB**.
∴ **AO = CO** and **BO = DO**.

Q.E.D.

Theorem 15

The straight lines which join the ends of two equal and parallel straight lines towards the same parts are themselves equal and parallel.

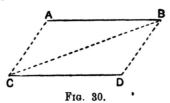

Fig. 30.

Given **AB** is equal and parallel to **CD**.
 To Prove **AC** is equal and parallel to **CD**
 Join **BC**.

In the △s **ABC, DCB**
  AB = DC, given.
  BC is common.
  ∠ **ABC** = ∠ **DCB** alt. angles, **AB** being ∥ to **CD**.
  ∴ △**ABC** ≡ △**DCB**.
  ∴ **AC = DB** and ∠ **ACB** = ∠ **DBC**.
But these are alt. angles, ∴ **AC** is parallel to **DB**.

Q.E.D.

This theorem can also be stated as follows :—

A quadrilateral which has one pair of equal and parallel sides is a parallelogram.

Other tests for a parallelogram are :—

(1) If the diagonals of a quadrilateral bisect each other, it is a parallelogram.
(2) If the opposite sides of a quadrilateral are equal, it is a parallelogram.
(3) If the opposite angles of a quadrilateral are equal, it is a parallelogram.

For riders on Theorems 11, 12, 13, 14, 15, see page 23.

BOOK II

Theorem 16

(1) Parallelograms on the same base and between the same parallels are equal in area.
(2) The area of a parallelogram is measured by the product of its base and its height.

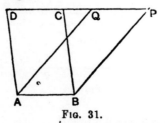

Fig. 31.

(1) *Given* **ABCD, ABPQ** are two parallelograms on the same base **AB** and between the same parallels **AB, DP**.

PROOFS OF THEOREMS

To Prove that **ABCD, ABPQ** are equal in area.

In the △s **AQD, BPC,**

∠ **ADQ** = ∠ **BCP**, corresp. ∠ s; **AD, BC** being ∥ lines.
∠ **AQD** = ∠ **BPC**, corresp. ∠ s; **AQ, BP** being ∥ lines.
AD = **BC**, opp. sides ∥gram.

∴ △**AQD** ≡ △**BPC**.

From the figure **ABPD**, subtract in succession each of the equal triangles **BPC, AQD**.

∴ the remaining figures **ABCD, ABPQ** are equal in area.

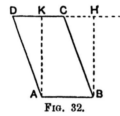

Fig. 32.

(2) If **BH** is the perpendicular from **B** to **CD**, the area of **ABCD** is measured by **AB . BH**.

Complete the rectangle **ABHK**.

The ∥gram **ABCD** and the rectangle **ABHK** are on the same base and between the same parallels and are therefore equal in area.

But the area of **ABHK** = **AB . BH**;

∴ the area of **ABCD** = **AB . BH**.

Q.E.D.

Theorem 17

The area of a triangle is measured by half the product of the base and the height.

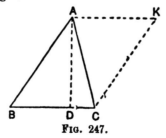

Fig. 247.

Given that **AD** is the perpendicular from **A** to the base **BC** of the triangle **ABC**.

To Prove that the area of △**ABC** = ½**AD** . **BC**.

Complete the parallelogram **ABCK**.

Since the diagonal **AC** bisects the parallelogram **ABCK**,

△**ABC** = ½ parallelogram **ABCK**.

But parallelogram **ABCK** = **AD** . **BC** ;

∴ △**ABC** = ½**AD** . **BC**.

Q.E.D.

Theorem 18

(1) Triangles on the same base and between the same parallels are equal in area.

(2) Triangles of equal area on the same base and on the same side of it are between the same parallels.

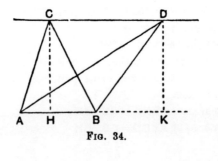

Fig. 34.

(1) *Given* two triangles **ABC**, **ABD** on the same base **AB** and between the same parallels **AB**, **CD**.

To Prove the triangles **ABC**, **ABD** are equal in area.

Draw **CH**, **DK** perpendicular to **AB** or **AB** produced.

△**CAB** = ½**CH** . **AB**.

△**DAB** = ½**DK** . **AB**.

But **CH** is parallel to **DK**, since each is perpendicular to **AB**, and **CD** is given parallel to **HK**.

∴ **CDKH** is a parallelogram.

∴ **CH** = **DK**, opp. sides.

∴ △**CAB** equals △**DAB** in area.

PROOFS OF THEOREMS 221

(2) *Given* two triangles **ABC**, **ABD** of equal area.
To Prove **CD** is parallel to **AB**.
Draw **CH**, **DK** perpendicular to **AB** or **AB** produced.
Now △**CAB** = ½**CH** . **AB** and △**DAB** = ½**DK** . **AB**.
∴ **CH** . **AB** = **DK** . **AB**.
∴ **CH** = **DK**.
But **CH** is parallel to **DK**, for each is perpendicular to **AB**.
∴ Since **CH** and **DK** are equal and parallel, **CHKD** is a parallelogram.
∴ **CD** is parallel to **HK** or **AB**.

Q.E.D.

THEOREM 19

If a triangle and a parallelogram are on the same base and between the same parallels, the area of the triangle is equal to half that of the parallelogram.

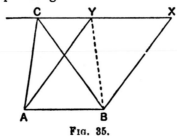

FIG. 35.

Given the triangle **ABC** and the parallelogram **ABXY** on the same base **AB** and between the same parallels **AB**, **CX**.
To Prove △**ABC** = ½ ∥gram **ABXY**.
Join **BY**.
The △s **ABC**, **ABY** are on the same base and between the same parallels.
∴ △**ABC** = △**ABY** in area.
Since the diagonal **BY** bisects the ∥gram **ABXY**,
△**ABY** = ½ ∥gram **ABXY**;
∴ △**ABC** = ½ ∥gram **ABXY**. Q.E.D.

The following formula for the area of a triangle is important:—
If a, b, c are the lengths of the sides of a triangle and if $s = \frac{1}{2}(a+b+c)$, the area of the triangle
$$= \sqrt{s(s-a)(s-b)(s-c)}.$$

By using the results :
   Area of parallelogram = height × base,
   Area of triangle   = ½ height × base.

Proofs similar to the proof of Theorem 18 can be easily obtained for the following theorems :—

(1) Triangles on equal bases and between the same parallels are equal in area.

(2) Parallelograms on equal bases and between the same parallels are equal in area.

(3) Triangles of equal area, which are on equal bases in the same straight line and on the same side of it, are between the same parallels.

(4) Parallelograms of equal area, which are on equal bases in the same straight line and on the same side of it, are between the same parallels.

(5) The area of a trapezium = the product of half the sum of the parallel sides and the distance between them.

For riders on Theorems 16, 17, 18, 19, see page 28.

THEOREM 20. [PYTHAGORAS' THEOREM.]

In any right-angled triangle, the square on the hypotenuse is equal to the sum of the squares on the sides containing the right angle.

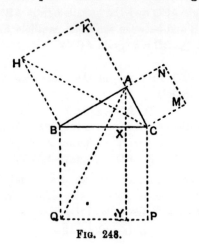

FIG. 248.

PROOFS OF THEOREMS   223

Given ∠ **BAC** is a right angle.

   To Prove the square on **BC** = the square on **BA** + the square on **AC**.

   Let **ABHK, ACMN, BCPQ** be the squares on **AB, AC, BC**.
   Join **CH, AQ**. Through **A**, draw **AXY** parallel to **BQ** cutting **BC, QP** at **X, Y**.
   Since ∠ **BAC** and ∠ **BAK** are right angles, **KA** and **AC** are in the same straight line.
   Again ∠ **HBA** = 90° = ∠ **QBC**.
   Add to each ∠ **ABC**, ∴ ∠ **HBC** = ∠ **ABQ**.
   In the △s **HBC, ABQ**

            HB = AB,   sides of square.
            CB = QB,   sides of square.
            ∠ **HBC** = ∠ **ABQ,** proved.
   ∴   △**HBC** ≡ △**ABQ**.

   Now △**HBC** and square **HA** are on the same base **HB** and between the same parallels **HB, KAC**;
   ∴   △**HBC** = ½ square **HA**.

   Also △**ABQ** and rectangle **BQYX** are on the same base **BQ** and between the same parallels **BQ, AXY**.
   ∴   △**ABQ** = ½ rect. **BQYX**.
   ∴   square **HA** = rect. **BQYX**.

   Similarly, by joining **AP, BM**, it can be shown that square **MA** = rect. **CPYX**;
   ∴   square **HA** + square **MA** = rect. **BQYX** + rect. **CPYX**
                                     = square **BP**.

                                                 Q.E.D.

Theorem 21

If the square on one side of a triangle is equal to the sum of the squares on the other sides, then the angle contained by these sides is a right angle.

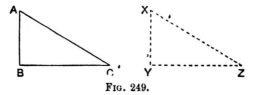

Fig. 249.

Given $AB^2 + BC^2 = AC^2$.

To Prove $\angle ABC = 90°$.

Construct a triangle XYZ such that XY = AB, YZ = BC, $\angle XYZ = 90°$.

Since $\angle XYZ = 90°$, $XZ^2 = XY^2 + YZ^2$.

But XY = AB and YZ = BC.

∴ $XZ^2 = AB^2 + BC^2 = AC^2$ given.

∴ XZ = AC.

∴ in the △s ABC, XYZ

AB = XY, constr.
BC = YZ, constr.
AC = XZ, proved.

∴ △ABC ≡ △XYZ.
∴ ∠ABC = ∠XYZ.

But $\angle XYZ = 90°$ constr.
∴ $\angle ABC = 90°$.

Q.E.D.

For riders on Theorems 20, 21, see page 38.

DEFINITION.—If AB and CD are any two straight lines, and if AH, BK are the perpendiculars from A, B to CD, then HK is called the *projection* of AB on CD.

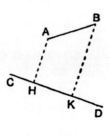

FIG. 250(1).

Thus, in Fig. 248,
QY is the projection of BA on QP,
XC is the projection of AC on BC,
BX is the projection of QA on BC.

Or, in Fig. 250(2),
AN is the projection of AC on AB,
BN is the projection of BC on AB.

Theorem 22

In an obtuse-angled triangle, the square on the side opposite the *obtuse* angle is equal to the sum of the squares on the sides containing it *plus* twice the rectangle contained by one of those sides and the projection on it of the other.

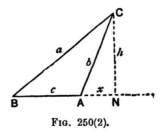

Fig. 250(2).

Given ∠ BAC is obtuse and CN is the perpendicular from C to BA produced.

To Prove $BC^2 = BA^2 + AC^2 + 2BA \cdot AN$.

[Put in a small letter for each length that comes in the answer and also for the altitude.]

Let $BC = a$ units, $BA = c$ units, $AC = b$ units, $AN = x$ units, $CN = h$ units.

It is required to prove that $a^2 = c^2 + b^2 + 2cx$.

Since ∠ BNC = 90°, $a^2 = (c + x)^2 + h^2$,

∴ $a^2 = c^2 + 2cx + x^2 + h^2$.

Since ∠ ANC = 90°, $b^2 = x^2 + h^2$,

∴ $a^2 = c^2 + 2cx + b^2$,

or $BC^2 = BA^2 + AC^2 + 2BA \cdot AN$.

Q.E.D.

Theorem 23

In any triangle, the square on the side opposite an *acute* angle is equal to the sum of the squares on the sides containing it *minus* twice the rectangle contained by one of those sides and the projection on it of the other.

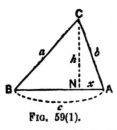
Fig. 59(1).

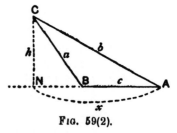
Fig. 59(2).

Given ∠ **BAC** is acute and **CN** is the perpendicular from **C** to **AB** or **AB** produced.

To Prove **BC² = BA² + AC² − 2AB . AN**.

[Put in a small letter for each length that comes in the answer and also for the height.]

Let **BC** = a units, **BA** = c units, **AC** = b units, **AN** = x units, **CN** = h units.

It is required to prove that $a^2 = c^2 + b^2 - 2cx$.

In Fig. 59(1), **BN** = $c - x$; in Fig. 59(2), **BN** = $x - c$.

Since ∠ **CNB** = 90°, $a^2 = (c-x)^2 + h^2$ in Fig. 59(1),
 or $a^2 = (x-c)^2 + h^2$ in Fig. 59(2);

∴ in each case, $a^2 = c^2 - 2cx + x^2 + h^2$.

Since ∠ **ANC** = 90°, $b^2 = x^2 + h^2$,

∴ $a^2 = c^2 - 2cx + b^2$,

or **BC² = BA² + AC² − 2 AB . AN**.

Q.E.D.

Theorem 24. [Apollonius' Theorem.]

In any triangle, the sum of the squares on two sides is equal to twice the square on half the base *plus* twice the square on the median which bisects the base.

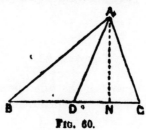
Fig. 60.

PROOFS OF THEOREMS 227

Given D is the mid-point of BC.
To Prove $AB^2 + AC^2 = 2AD^2 + 2BD^2$.
Draw AN perpendicular to BC.
From the triangle ADB, $\quad AB^2 = AD^2 + DB^2$
From the triangle ADC, $\quad AC^2 = AD^2 + DC^2 - 2DC \cdot DN$.
But $BD = DC$, given; $\therefore BD \cdot DN = DC \cdot DN$ and $BD^2 = DC^2$
$\therefore$ adding, $\quad AB^2 + AC^2 = 2AD^2 + 2DB^2$.

Q.E.D.

For riders on Theorems 22, 23, 24, see page 44.

Theorem 25

(1) If A, B, C, D are four points in order on a straight line, then
$AC \cdot BD = AB \cdot CD + AD \cdot BC$.
(2) If a straight line AB is bisected at O, and if P is any other
point on AB, then $AP^2 + PB^2 = 2AO^2 + 2OP^2$.

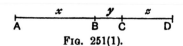
Fig. 251(1).

(1) Let $AB = x$ units, $BC = y$ units, $CD = z$ units.
Then $\quad AC = x + y$, $BD = y + z$.
$\therefore AC \cdot BD = (x + y)(y + z)$
$\qquad = xy + y^2 + xz + yz$.
Also $AD = x + y + z$.
$\therefore AB \cdot CD + AD \cdot BC = xz + (x + y + z)y$
$\qquad = xz + xy + y^2 + yz$.
$\therefore AC \cdot BD = AB \cdot CD + AD \cdot BC$.

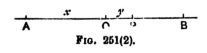
Fig. 251(2).

(2) Let $AO = x$ units, $OP = y$ units.
$\therefore OB = AO = x$.
Also $PB = OB - OP = x - y$
and $AP = AO + OP = x + y$.

$$\therefore \mathsf{AP}^2 + \mathsf{PB}^2 = (x+y)^2 + (x-y)^2$$
$$= x^2 + 2xy + y^2 + x^2 - 2xy + y^2$$
$$= 2x^2 + 2y^2$$
$$= 2\mathsf{AO}^2 + 2\mathsf{OP}^2.$$

Q.E.D.

For riders on Theorem 25, see page 46.

GEOMETRICAL ILLUSTRATIONS OF ALGEBRAIC IDENTITIES

I $\qquad (a+b)^2 = a^2 + 2ab + b^2.$

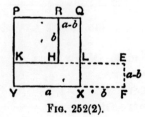

Fig. 252(1).

Draw a line **PQ** of length $a+b$ inches and take a point **R** on it such that **RQ** is of length b inches.

On **PQ** and **RQ** describe squares **PQXY**, **RQHK** on the same side of **PQ** and produce **RK**, **HK** to meet **XY**, **PY** at **M**, **L**.

Then the area of **PQXY** is $(a+b)^2$ sq. inches.

The areas of **LKMY** and **RQHK** are a^2 sq. inches and b^2 sq. inches.

The area of each of the rectangles **PK**, **KX** is ab sq. inches.

$$\therefore \quad (a+b)^2 = a^2 + 2ab + b^2.$$

II. $\qquad (a+b)(a-b) = a^2 - b^2.$

Fig. 252(2).

PROOFS OF THEOREMS

Draw a line **PQ** of length a inches ($a > b$) and cut off a part **PR** of length b inches.

On **PQ** and **PR** describe squares **PQXY**, **PRHK**; produce **KH** to meet **QX** at **L**.

Produce **KL**, **YX** to **E**, **F** so that **LE** = **XF** = b inches.

Now **LX** = **QX** − **QL** = **QX** − **RH** = $a - b$ inches.

∴ the rectangle **LXFE** equals the rectangle **HLQR**.

∴ the rectangle **KYFE** equals the sum of the rectangles **KYXL** and **HLQR** equals **PQXY** − **PRHK** = $a^2 - b^2$ sq. in.

But **KY** = $a - b$ inches, **YF** = $a + b$ inches.

$$\therefore (a+b)(a-b) = a^2 - b^2.$$

Theorem 26

(1) If two sides of a triangle are unequal, the greater side has the greater angle opposite to it.

(2) If two angles of a triangle are unequal, the greater angle has the greater side opposite to it.

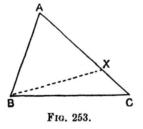

Fig. 253.

(1) *Given* **AC** > **AB**

To Prove ∠ **ABC** > ∠ **ACB**.

From **AC** cut off a part **AX** equal to **AB**. Join **BX**.

Since **AB** = **AX**, ∠ **ABX** = ∠ **AXB**.

But ext. ∠ **AXB** > int. opp. ∠ **XCB**,

∴ ∠ **ABX** > ∠ **XCB**.

But ∠ **ABC** > ∠ **ABX**,

∴ ∠ **ABC** > ∠ **XCB** or ∠ **ACB**.

(2) *Given* ∠ ABC > ∠ ACB.
   To Prove AC > AB.
   If AC is not greater than AB, it must either be equal to AB, or less than AB.

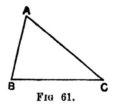

Fig 61.

If AC = AB, ∠ ABC = ∠ ACB, which is contrary to hypothesis.
If AC < AB, ∠ ABC < ∠ ACB, which is contrary to hypothesis.
   ∴ AC must be greater than AB.

Q.E.D.

Theorem 27

Of all straight lines that can be drawn to a given straight line from an external point, the perpendicular is the shortest.

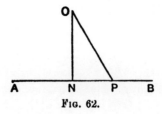

Fig. 62.

Given a fixed point O and a fixed line AB.
   ON is the perpendicular from O to AB, and OP is any other line from O to AB.
   To Prove ON < OP.
   Since the sum of the angles of a triangle is 2 rt. angles, and since ∠ ONP = 1 rt. angle.
      ∴ ∠ NPO + ∠ NOP = 1 rt. angle.
      ∴ ∠ NPO < 1 rt. angle.
      ∴ ∠ NPO < ∠ ONP.
      ∴ ON < OP.

Q.E.D

THEOREM 28

Any two sides of a triangle are together greater than the third side.

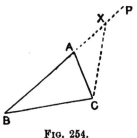

FIG. 254.

Given the triangle ABC.

To Prove BA + AC > BC.

Produce BA to P and cut off AX equal to AC. Join CX.

Since AX = AC, ∠ACX = ∠AXC.

But ∠BCX > ∠ACX.

∴ ∠BCX > ∠AXC.

∴ in the triangle BXC, ∠BCX > ∠BXC.

∴ BX > BC.

But BX = BA + AX = BA + AC.

∴ BA + AC > BC.

Q.E.D.

The following theorem is an easy rider on the above :—

The shortest and longest distances from a point to a circle lie along the diameter through the point.

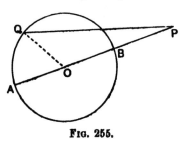

FIG. 255.

If AB is a diameter, and if P lies on AB produced, PA > PQ > PB.

Join Q to the centre O.
$$PA = PO + OA = PO + OQ > PQ.$$
$$PB + BO = PO < PQ + QO.$$
For riders on Theorems 26, 27, 28 see page 49.

Theorem 29

The straight line joining the middle points of two sides of a triangle is parallel to the base and equal to half the base.

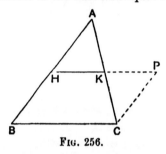

Fig. 256.

Given H, K are the middle points of AB, AC.

To Prove HK is parallel to BC and HK = ½ BC.

Through C, draw CP parallel to BA to meet HK produced at P.

In the △s AHK, CPK.
$$\angle AHK = \angle CPK, \text{ alt. } \angle \text{s.}$$
$$\angle HAK = \angle PCK, \text{ alt. } \angle \text{s.}$$
$$AK = KC, \text{ given.}$$
∴ △AHK ≡ △CPK.
∴ CP = AH.
But AH = BH, given.
∴ CP = BH.
Also CP is drawn parallel to BH.
∴ the lines CP, BH are equal and parallel.
∴ BCPH is a parallelogram.
∴ HK is parallel to BC.
Also HK = KP from congruent triangles.
∴ HK = ½HP.
But HP = BC opp. sides of parallelogram.
∴ HK = ½BC.   Q.E.D.

PROOFS OF THEOREMS 233

Theorem 30

If there are three or more parallel straight lines, and if the intercepts made by them on any straight line cutting them are equal, then the intercepts made by them on any other straight line that cuts them are equal.

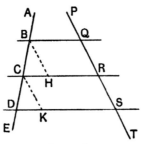

Fig. 257.

Given three parallel lines cutting a line **AE** at **B, C, D** and any other line **PT** at **Q, R, S** and that **BC = CD**.

To Prove **QR = RS**.

Draw **BH, CK** parallel to **PT** to meet **CR, DS** at **H, K**.
Then **BH** is parallel to **CK**.
∴ in the △s **BCH, CDK**.
$\qquad$ ∠ **CBH** = ∠ **DCK** corresp. ∠s.
$\qquad$ ∠ **BCH** = ∠ **CDK** corresp. ∠s.
$\qquad$ **BC = CD**, given.
∴ △**BCH** ≡ △**CDK**.
∴ **BH = CK**.
But **BQRH** is a ∥gram since its opposite sides are parallel.
$\qquad$ ∴ **BH = QR**.
And **CRSK** is a ∥gram since its opposite sides are parallel.
$\qquad$ ∴ **CK = RS**.
$\qquad$ ∴ **QR = RS**. •

Q.E.D.

For riders on Theorems 29, 30 see page 52.

BOOK III

Theorem 31

(1) The straight line which joins the centre of a circle to the middle point of a chord (which is not a diameter) is perpendicular to the chord.

(2) The line drawn from the centre of a circle perpendicular to a chord bisects the chord.

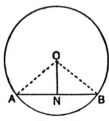

Fig. 69.

(1) *Given* a circle, centre **O**, and a chord **AB**, whose mid-point is **N**.

To Prove ∠ **ONA** is a right angle.

Join **OA, OB**.

In the △s **ONA, ONB**,

        OA = **OB**, radii.
        AN = **BN**, given.
        ON is common.

∴ △ **ONA** ≡ △ **ONB**.
∴ ∠ **ONA** = ∠ **ONB**.

But these are adjacent angles, ∴ each is a right angle.

(2) *Given* that **ON** is the perpendicular from the centre **O** of a circle to a chord **AB**.

To Prove that **N** is the mid-point of **AB**.

In the *right-angled* triangles **ONA, ONB**.

        OA = **OB**, radii.
        ON is common.

∴ △ **ONA** ≡ △ **ONB**.
∴ **AN** = **NB**.

Q.E.D.

THEOREM 32

In equal circles or in the same circle :
  (1) Equal chords are equidistant from the centres.
  (2) Chords which are equidistant from the centres are equal.

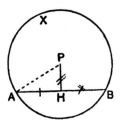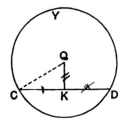

FIG. 258(1).

(1) *Given* two equal circles **ABX, CDY**, centres **P, Q**, and two equal chords **AB, CD**.

To Prove that the perpendiculars **PH, QK** from **P, Q** to **AB, CD** are equal.

Join **PA, QC**.

Since **PH, QK** are the perpendiculars from the centres to the chords **AB, CD**, **H** and **K** are the mid-points of **AB** and **CD**.

∴ $AH = \tfrac{1}{2}AB$ and $CK = \tfrac{1}{2}CD$.

But **AB = CD**, given.

∴ **AH = CK**.

∴ in the *right-angled* triangles **PAH, QCK**, the hypotenuse **PA** = the hypotenuse **QC**, radii of equal circles.

AH = CK, proved.

∴ △ **PAH** ≡ △ **QCK**.

∴ **PH = QK**.

Q.E.D.

(2) *Given* that the perpendiculars **PH, QK** from **P, Q** to the chords **AB, CD** are equal.

To Prove that **AB = CD**.

In the *right-angled* triangles **PAH, QCK**, the hypotenuse **PA** = the hypotenuse **QC**, radii of equal circles.

PH = QK, given.

∴ △ **PAH** ≡ △ **QCK**.

∴ **AH = CK**.

236   CONCISE GEOMETRY

But the perpendiculars **PH, QK** bisect **AB, CD**.
∴   **AB** = 2**AH** and **CD** = 2**CK**.
∴   **AB** = **CD**.

Q.E.D.

The proof is unaltered if the chords are in the same circle.

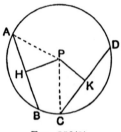

Fig. 258(2).

For riders on Theorems 31, 32, see page 57.

Theorem 33

The angle which an arc of a circle subtends at the centre is double that which it subtends at any point on the remaining part of the circumference.

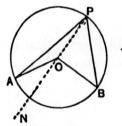 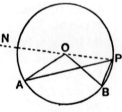 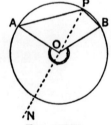

Fig. 259(1).   Fig. 259(2).   Fig. 259(3).

Given **AB** is an arc of a circle, centre **O**; **P** is any point on the remaining part of the circumference.

To Prove   ∠ **AOB** = 2 ∠ **APB**.

Join **PO**, and produce it to any point **N**.
Since   **OA** = ⊙**P**, ∠ **OAP** = ∠ **OPA**.
But   ext. ∠ **NOA** = int. ∠ **OAP** + int. ∠ **OPA**.
∴   ∠ **NOA** = 2 ∠ **OPA**.

PROOFS OF THEOREMS 237

Similarly ∠ NOB = 2 ∠ OPB.
∴ adding in Fig. 259(1) and subtracting in Fig. 259(2), we have   ∠ AOB = 2 ∠ APB.

Q.E.D.

Fig. 259(3) shows the case where the angle AOB is reflex, *i.e.* greater than 180°: the proof for Fig. 259(3) is the same as for Fig. 259(1).

THEOREM 34

1) Angles in the same segment of a circle are equal.
2) The angle in a semicircle is a right angle.

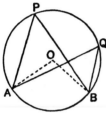   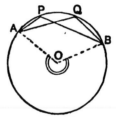

FIG. 76(1).   FIG. 76(2).

1) *Given* two angles APB, AQB in the same segment of a circle.
   To Prove ∠ APB = ∠ AQB.
   Let O be the centre. Join OA, OB.
   Then ∠ AOB = 2 ∠ APB.   ∠ at centre = twice ∠ at ○ce.
   and   ∠ AOB = 2 ∠ AQB.
   ∴   ∠ APB = ∠ AQB.

Q.E.D.

2) *Given* AB a diameter of a circle, centre O, and P a point on the circumference.
   To Prove ∠ APB = 90°.

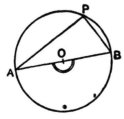

FIG. 77.

∠ AOB = 2 ∠ APB.   ∠ at centre = twice ∠ at ○ce.
But ∠ AOB = 180°, since AOB is a straight line;
∴ ∠ APB = 90°.

Q.E.D.

Theorem 35

(1) The opposite angles of a cyclic quadrilateral are supplementary.
(2) If a side of a cyclic quadrilateral is produced, the exterior angle is equal to the interior opposite angle.

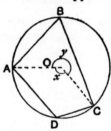

Fig. 260.

(1) *Given* ABCD is a cyclic quadrilateral.
   To Prove ∠ ABC + ∠ ADC = 180°.
   Let O be the centre of the circle. Join OA, OC.
   Let the arc ADC subtend angle $x°$ at the centre,
      and let the arc ABC subtend angle $y°$ at the centre,
         ∴ $x° + y° = 360°$.
   Now $x° = 2$ ∠ ABC.   ∠ at centre = twice ∠ at ○ce.
   and $y° = 2$ ∠ ADC.
   ∴ 2 ∠ ABC + 2 ∠ ADC = 360°.
   ∴ ∠ ABC + ∠ ADC = 180°.

Q.E.D.

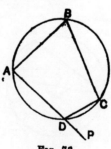

Fig. 78.

(2) *Given* the side AD of the cyclic quadrilateral ABCD is produced to P.
  To Prove   ∠ PDC = ∠ ABC.
  Now ∠ ADC + ∠ PDC = 180°, adj. angles.
  and ∠ ADC + ∠ ABC = 180°, opp. ∠ s cyclic quad.
  ∴   ∠ ADC + ∠ PDC = ∠ ADC + ∠ ABC.
  ∴   ∠ PDC = ∠ ABC.   Q.E.D.

For riders on Theorems 33, 34, 35 see page 62.

THEOREM 36

(1) If the line joining two points subtends equal angles at two other points on the same side of it, then the four points lie on a circle.
(2) If the opposite angles of a quadrilateral are supplementary, then the quadrilateral is cyclic.

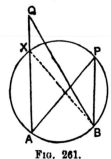

FIG. 261.

(1) *Given* that ∠ APB = ∠ AQB where P, Q are points on the same side of AB.
  To Prove that A, P, Q, B lie on a circle.
  If possible, let the circle through A, B, P not pass through Q and let it cut AQ or AQ produced at X. Join BX.
  Then ∠ AXB = ∠ APB, same segment,
  and   ∠ AQB = ∠ APB, given.
  ∴   ∠ AXB = ∠ AQB.
  that is, the exterior angle of the triangle BQX equals the interior opposite angle, which is impossible.
  ∴ the circle through A, B, P must pass through Q.
                                                        Q.E.D.

(2) *Given* that in the quadrilateral **ABCD**, ∠ **ABC** + ∠ **ADC** = 180°.
To Prove that **A, B, C, D** lie on a circle.

If possible let the circle through **A, B, C** not pass through **D**, and let it cut **AD** or **AD** produced at **X**. Join **CX**.

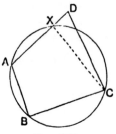

Fig. 262.

Then ∠ **ABC** + ∠ **AXC** = 180°, opp. ∠ s cyclic quad.
But ∠ **ABC** + ∠ **ADC** = 180°, given.
∴ ∠ **AXC** = ∠ **ADC**.

That this, the exterior angle of the triangle **CXD** equals the interior opposite angle, which is impossible.

∴ the circle through **A, B, C** must pass through **D**.

Q.E.D.

For riders on Theorem 36, see page 83.

Theorem 37

In equal circles (or in the same circle), if two arcs subtend equal angles at the centres or at the circumferences, they are equal.

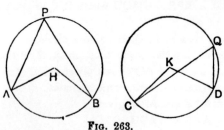

Fig. 263.

Given two equal circles, **ABP, CDQ**, centres **H, K**.
(1) *Given* that ∠ **AHB** = ∠ **CKD**.
To Prove that arc **AB** = arc **CD**

PROOFS OF THEOREMS

Apply the circle **AB** to the circle **CD** so that the centre **H** falls on the centre **K** and **HA** along **KC**.

Since the circles are equal, **A** falls on **C** and the circumferences coincide.

Since $\angle$ **AHB** = $\angle$ **CKD**, **HB** falls on **KD**, and **B** falls on **D**.

∴ the arcs **AB**, **CD** coincide.

∴ arc **AB** = arc **CD**.

(2) *Given* that $\angle$ **APB** = $\angle$ **CQD**.

To Prove that arc **AB** = arc **CD**.

Now $\angle$ **AHB** = 2 $\angle$ **APB**, $\angle$ at centre = twice $\angle$ at O̊ce.

and $\angle$ **CKD** = 2 $\angle$ **CQD**.

But $\angle$ **APB** = $\angle$ **CQD**, given.

∴ $\angle$ **AHB** = $\angle$ **CKD**.

∴ arc **AB** = arc **CD**.

Q.E.D.

Theorem 38

In equal circles (or in the same circle), if two arcs are equal, they subtend equal angles at the centres and at the circumferences.

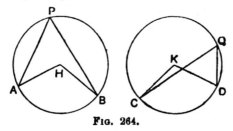

Fig. 264.

Given two equal circles **ABP**, **CDQ**, centres **H**, **K**, and two equal arcs **AB**, **CD**.

To Prove (1) $\angle$ **AHB** = $\angle$ **CKD**.

(2) $\angle$ **APB** = $\angle$ **CQD**.

(1) Apply the circle **AB** to the circle **CD** so that the centre **H** falls on the centre **K** and **HA** along **KC**.

Since the circles are equal, **A** falls on **C** and the circumferences coincide.

But arc **AB** = arc **CD**, ∴ **B** falls on **D** and **HB** on **KD**.

∴ $\angle$ **AHB** coincides with $\angle$ **CKD**.

∴ $\angle$ **AHB** = $\angle$ **CKD**.

Q.E.D.

242   CONCISE GEOMETRY

(2) Now ∠APB = ½ ∠AHB.   ∠at ○ce = ½ ∠ at ..
  ∠CQD = ½ ∠CKD.
But ∠AHB = ∠CKD, just proved.
∴ ∠APB = ∠CQD.

Q.E.D.

THEOREM 39

In equal circles or in the same circle
  (1) if two chords are equal, the arcs which they cut off are equal
  (2) if two arcs are equal, the chords of those arcs are equal.

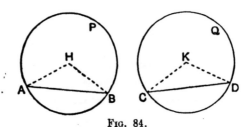

Fig. 84.

Given two equal circles ABP, CDQ, centres H, K.
(1) *Given* chord AB = chord CD.
  To Prove arc AB = arc CD.
  Join HA, HB, KC, KD.
  In the △s HAB, KCD,
    HA = KC, radii of equal circles.
    HB = KD, radii of equal circles.
    AB = CD, given.
  ∴ △HAB ≡ △KCD.
  ∴ ∠AHB = ∠CKD.
  ∴ the arcs AB, CD of equal circles subtend equal angles at the centres.
    ∴ arc AB = arc CD.

Q.E.D.

(2) *Given* arc AB = arc CD.
  To Prove chord AB = chord CD.
  Since AB, CD are equal arcs of equal circles,
    ∠AHB = ∠CKD.

PROOFS OF THEOREMS

∴ in the △s HAB, KCD,
        HA = KC, radii of equal circles.
        HB = KD, radii of equal circles.
    ∠ AHB = ∠ CKD, proved.
∴   △HAB ≡ △KCD.
∴   AB = CD.

Q.E.D.

For riders on Theorems 37, 38, 39 see page 72.

THE TANGENT TO A CIRCLE

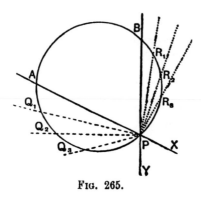

Fig. 265.

Let P be any point on an arc AB of a circle.

Suppose a point Q starts at A and moves along the arc AP towards P, taking successive positions $Q_1, Q_2, Q_3 \ldots$ and draw the lines $PQ_1, PQ_2, PQ_3 \ldots$

Also suppose a point R starts at B and moves along the arc BP towards P, taking successive positions $R_1, R_2, R_3 \ldots$ and draw the lines $PR_1, PR_2, PR_3 \ldots$

All lines in the PQ system cut off arcs along PA, the lengths of which decrease without limit as Q tends to P.

All lines in the PR system cut off arcs along PB, the lengths of which decrease without limit as R tends to P.

Produce AP, BP to X, Y.

All lines drawn from P in the angle APY or BPX belong either to the **PQ** system or to the **PR** system, except the single line which cuts off an arc of zero length.

This line is called the *tangent at* **P**.

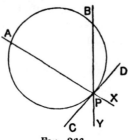

Fig. 266.

The tangent at **P** is therefore the line **CPD**, which is the intermediate position between lines of the **PQ** system and lines of the **PR** system, and cuts off an arc of zero length at **P**.

Theorem 40

The tangent to a circle is at right angles to the radius through the point of contact.

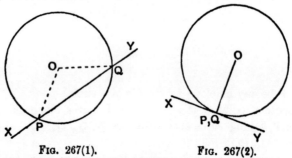

Fig. 267(1).   Fig. 267(2).

Given **P** is any point on a circle, centre **O**.

To Prove the tangent at **P** is perpendicular to **OP**.

Through P, see Fig. 267(1), draw any line **XPQY**, cutting the circle again at Q. Join OP, OQ.

$\quad$ OP = OQ, radii,

∴ ∠OPQ = ∠OQP.

∴ their supplements are equal,

∴ ∠OPX = ∠OQY.

PROOFS OF THEOREMS   245

Now the tangent at P is the limiting position of the line
XPQY, when the arc PQ is decreased without limit, so
that Q coincides with P, see Fig. 267(2).
∴ in Fig. 267(2), ∠OPX = ∠OPY.
But these are adjacent angles, ∴ each is a right angle.
∴ in Fig. 267(2), ∠OPX = 90°, where PX is the tangent at P.
Q.E.D.

THEOREM 41

If a straight line touches a circle and, from the point of contact,
a chord is drawn, the angles which the chord makes with
the tangent are equal to the angles in the alternate segments
of the circle.

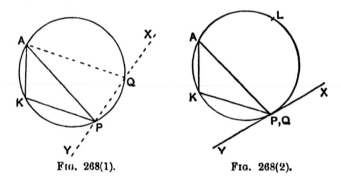

FIG. 268(1).   FIG. 268(2).

Given YPX is a tangent at P to the circle PLAK, and PA is any
chord through P.
To Prove ∠APX = ∠PKA and ∠APY = ∠PLA.
In Fig. 268(1), draw through P any line YPQX cutting the
circle again at Q. Join QA.
Then ∠AQX = ∠PKA; ext. ∠ of cyclic quad. = int. opp. ∠.
Now the tangent at P is the limiting position of the line
YPQX when the arc PQ is decreased without limit, so
that Q coincides with P, see Fig. 268(2).
But the limiting position of ∠AQX is ∠APX.
∴ when YPQX becomes the tangent at P,
∠APX = ∠PKA.
Similarly it may be proved that ∠APY = ∠PLA.
Q.E.D.

246 CONCISE GEOMETRY

The converse of this theorem is frequently of use in rider-work. For riders on Theorems 40, 41, see page 68.

Theorem 42

If two tangents are drawn to a circle from an external point—
  (1) The tangents are equal.
  (2) The tangents subtend equal angles at the centre.
  (3) The line joining the centre to the external point bisects the angle between the tangents.

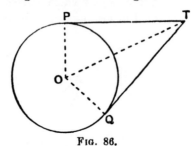

Fig. 86.

Given **TP, TQ** are the tangents from **T** to a circle, centre **O**.
  To Prove (1) **TP = TQ**.
    (2) ∠**TOP** = ∠**TOQ**.
    (3) ∠**OTP** = ∠**OTQ**.
  Since **TP, TQ** are tangents at **P, Q**, the angles **TPO, TQO** are right angles.
  ∴ in the *right-angled* triangles **TOP, TOQ**
    OP = OQ, radii.
    OT is the common hypotenuse.
    ∴ △**TOP** ≡ △**TOQ**.
    ∴   **TP = TQ**,
    and ∠**TOP** = ∠**TOQ**,
    and ∠**OTP** = ∠**OTQ**.

Q.E.D.

Theorem 43

If two circles touch one another, the line joining their centres (produced if necessary) passes through the point of contact.

PROOFS OF THEOREMS 247

Given two circles, centres **A, B**, touching each other at **P**.
To Prove **AB** (produced if necessary) passes through **P**.

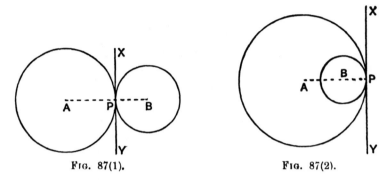

Fig. 87(1).   Fig. 87(2).

Since the circles touch each other at **P**, they have a common tangent **XPY** at **P**.

Since **XP** touches each circle at **P**, the angles **X**∩**A**, **XPB** are right angles.

∴ **A** and **B** each lie on the line through **P** perpendicular to **PX**.

∴ **A, B, P** lie on a straight line.

Q.E.D.

Note.—If two circles touch each other externally (Fig. 87(1)), the distance between their centres equals the *sum* of the radii.

If two circles touch each other internally (Fig. 87(2)), the distance between their centres equals the *difference* of the radii.

For riders on Theorems 42, 43, see page 77.

Theorem 44

In a right-angled triangle, the line joining the mid-point of the hypotenuse to the opposite vertex is equal to half the hypotenuse.

Given **ABC** is a triangle, right-angled at **A**, and **D** is the mid-point of **BC**.

To Prove **AD** = ½**BC**.
Draw a circle through **A, B, C**.
Since ∠ **BAC** = 90°, **BC** is a diameter.

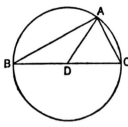

Fig. 269.

But **D** is the mid-point of **BC**, ∴ **D** is the centre of the circle.
∴   **DA** = **DB** = **DC**, radii.
∴   **DA** = ½**BC**.

Q.E.D.

Definition.—If a point moves in such a way that it obeys a given geometrical condition, the path traced out by the point is called the *locus* of the point.

Theorem 45

The locus of a point, which is equidistant from two given points, is the perpendicular bisector of the straight line joining the given points.

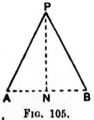

Fig. 105.

Given two fixed points **A, B** and any position of a point **P** which moves so that **PA** = **PB**.
To Prove that **P** lies on the perpendicular bisector of **AB**.
Bisect **AB** at **N**. Join **PN**.

PROOFS OF THEOREMS 249

In the △s ANP, BNP,
   AN = BN, constr.
   AP = BP, given.
   PN is common.
  ∴ △ANP ≡ △BNP.
  ∴ ∠ANP = ∠BNP.
But these are adjacent angles, ∴ each is a right angle.
 ∴ PN is perpendicular to AB and bisects it.
∴ P lies on the perpendicular bisector of AB.
Q.E.D.

Theorem 46

The locus of a point which is equidistant from two given intersecting straight lines is the pair of lines which bisect the angles between the given lines.

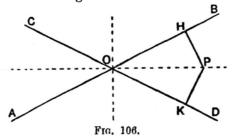

Fig. 106.

Given two fixed lines AOB, COD and any position of a point P which moves so that the perpendiculars PH, PK from P to AOB, COD are equal.

To Prove P lies on one of the two lines bisecting the angles BOC, BOD.

Suppose P is situated in the angle BOD.
In the *right-angled* triangles PHO, PKO,
   PH = PK, given.
   PO is the common hypotenuse.
  ∴ △PHO ≡ △PKO.
  ∴ ∠POH = ∠POK.
∴ P lies on the line bisecting the angle BOD.
In the same way if P is situated in either of the angles BOC, COA, AOD, it lies on the bisectors of these angles.

For riders on Theorems 45, 46, see page 94.   Q.E.D.

Theorem 47

The perpendicular bisectors of the three sides of a triangle are concurrent (*i.e.* meet in a point).

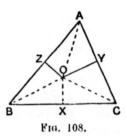

Fig. 108.

Given that the perpendicular bisectors **OY**, **OZ** of **AC**, **AB** meet at **O**.

To Prove the perpendicular bisector of **BC** passes through **O**.

Bisect **BC** at **X**, join **OX**; also join **OA**, **OB**, **OC**.

In the △s **OZA, OZB**,

   BZ = ZA, given.
   OZ is common.
   ∠ **BZO** = ∠ **AZO**, given rt. ∠s.

∴ △**OZA** = △**OZB**.
∴ **OA = OB**.

Similarly from the △s **OYA, OYC**, it can be proved that
   OA = OC,
∴ **OB = OC**.

In the △s **OXB, OXC**,

   OB = OC, proved.
   XB = XC, constr.
   OX is common.

∴ △**OXB** = △**OXC**.
∴ ∠ **OXB** = ∠ **OXC**.

But these are adjacent angles, ∴ each is a rt. ∠.

∴ **OX** is the perpendicular bisector of **BC**.

            Q.E.D.

For riders on Theorem 47, see page 99.

THEOREM 48

The internal bisectors of the three angles of a triangle are concurrent.

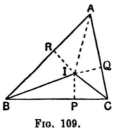

Fig. 109.

Given that the internal bisectors IB, IC of the angles ABC, ACB meet at I.

To Prove that IA bisects the angle BAC.

Join IA. Draw IP, IQ, IR perpendicular to BC, CA, AB.

In the △s IBP, IBR,
   ∠IBP = ∠IBR, given.
   ∠IPB = ∠IRB, constr. rt. ∠s.
   IB is common.
∴ △IBP ≡ △IBR.
∴ IP = IR.

Similarly from the △s ICP, ICQ it may be proved that
   IP = IQ,
∴ IQ = IR.

In the *right-angled* triangles IAQ, IAR,
   IQ = IR, proved.
   IA is the common hypotenuse.
∴ △IAQ ≡ △IAR.
∴ ∠IAQ = ∠IAR.
∴ IA bisects the angle BAC.

            Q.E.D.

For riders on Theorem 48, see page 190.

THEOREM 49

The three altitudes of a triangle (*i.e.* the lines drawn from the vertices perpendicular to the opposite sides) are concurrent.

Given **AD, BE, CF** are the altitudes of the triangle **ABC**.

To Prove **AD, BE, CF** are concurrent.

Through **A, B, C** draw lines parallel to **BC, CA, AB** to form the triangle **PQR**.

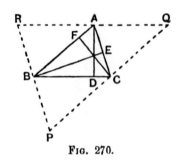

Fig. 270.

Since **BC** is ∥ to **AR** and **AC** is ∥ to **BR**,
  BCAR is a parallelogram.
 ∴ **BC = AR**.
Similarly, since **BCQA** is a parallelogram, **BC = AQ**,
 ∴ **AR = AQ**.
Since **AD** is perpendicular to **BC**, and since **QR, BC** are parallel,
 ∴ **AD** is perpendicular to **QR**.
But **AR = AQ**, ∴ **AD** is the perpendicular bisector of **QR**.
Similarly, **BE** and **CF** are the perpendicular bisectors of **PR, PQ**.
But the perpendicular bisectors of the sides of the triangle **PQR** are concurrent.
 ∴ **AD, BE, CF** are concurrent.

Q.E.D.

For riders on Theorem 49, see page 101.

Theorem 50

(1) The three medians of a triangle (*i.e.* the lines joining each vertex to the middle point of the opposite side) are concurrent.

PROOFS OF THEOREMS

(2) The point at which the medians intersect is one-third of the way up each median (measured towards the vertex).

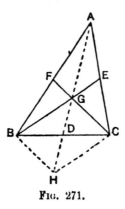

Fig. 271.

(1) *Given* the medians **BE, CF** of the triangle **ABC**, intersect at **G**.

To Prove that **AG**, when produced, bisects **BC**.
Join **AG** and produce it to **H**, so that **AG** = **GH**.
Let **AH** cut **BC** at **D**; join **HB, HC**.
Since **AF** = **FB** and **AG** = **GH**,
  FG is parallel to **BH**.
Since **AE** = **EC** and **AG** = **GH**,
  EG is parallel to **CH**.
Since **FGC** and **EGB** are parallel to **BH** and **CH**,
  BGCH is a parallelogram;
∴ the diagonals **BC, GH** bisect each other;
∴ **BD** = **DC**.

Q.E.D.

(2) For the same reason, **GD** = **DH**.
∴ **GH** = 2**GD**.
But **AG** = **GH**.
∴ **AG** = 2**GD**.
∴ **AD** = 3**GD**.
or **GD** = $\tfrac{1}{3}$**AD**.

Q.E.D.

For riders on Theorem 50, see page 103.

BOOK IV

Theorem 51

If two triangles have equal heights, the ratio of their areas is equal to the ratio of their bases.

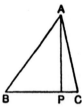 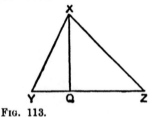

Fig. 113.

Given two triangles **ABC**, **XYZ** having equal heights **AP**, **XQ**.

To Prove $\dfrac{\triangle ABC}{\triangle XYZ} = \dfrac{BC}{YZ}$.

The area of a triangle $= \frac{1}{2}$ height × base.

$\therefore \triangle ABC = \frac{1}{2} AP \cdot BC$,

and $\triangle XYZ = \frac{1}{2} XQ \cdot YZ$,

$\therefore \dfrac{\triangle ABC}{\triangle XYZ} = \dfrac{\frac{1}{2} AP \cdot BC}{\frac{1}{2} XQ \cdot YZ}$.

But **AP** = **XQ**, given,

$\therefore \dfrac{\triangle ABC}{\triangle XYZ} = \dfrac{BC}{YZ}$.

Q.E.D.

Theorem 52

(1) If a straight line is drawn parallel to one side of a triangle, it divides the other sides (produced if necessary) proportionally.

(2) If a straight line divides two sides of a triangle proportionally, it is parallel to the third side.

 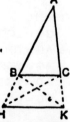 

Fig. 114(1).   Fig. 114(2).   Fig. 114(3).

PROOFS OF THEOREMS

(1) *Given* a line parallel to **BC** cuts **AB, AC** (produced if necessary) at **H, K**.

To Prove $\dfrac{AH}{HB} = \dfrac{AK}{KC}$.

Join **BK, CH**.

The triangles **KHA, KHB** have a common altitude from **K** to **AB**.

$$\therefore \frac{\triangle KHA}{\triangle KHB} = \frac{AH}{HB}.$$

The triangles **HKA, HKC** have a common altitude from **H** to **AC**.

$$\therefore \frac{\triangle HKA}{\triangle HKC} = \frac{AK}{KC}.$$

But $\triangle$ **KHB**, $\triangle$ **KHC** are equal in area, being on the same base **HK** and between the same parallels **HK, BC**.

$$\therefore \frac{AH}{HB} = \frac{AK}{KC}.$$

(2) *Given* a line **HK** cutting **AB, AC** at **H, K** such that $\dfrac{AH}{HB} = \dfrac{AK}{KC}$.

To Prove **HK** is parallel to **BC**.

The triangles **KHA, KHB** have a common altitude from **K** to **AB**.

$$\therefore \frac{\triangle KHA}{\triangle KHB} = \frac{AH}{HB}.$$

The triangles **HKA, HKC** have a common altitude from **H** to **AC**.

$$\therefore \frac{\triangle HKA}{\triangle HKC} = \frac{AK}{KC}.$$

But $\dfrac{AH}{HB} = \dfrac{AK}{KC}$, given.

$$\therefore \frac{\triangle KHA}{\triangle KHB} = \frac{\triangle HKA}{\triangle HKC}.$$

$\therefore \triangle KHB = \triangle HKC$.

But these triangles are on the same base **HK** and on the same side of it.

$\therefore$ **HK** is parallel to **BC**.

Q. E. D.

256 CONCISE GEOMETRY

COROLLARY 1.—If a line HK cuts AB, AC at H, K so that
$$\frac{AH}{HB} = \frac{AK}{KC},$$

Then $\frac{AH}{AB} = \frac{AK}{AC}$ and $\frac{HB}{AB} = \frac{KC}{AC}.$

Now $1 + \frac{AH}{HB} = 1 + \frac{AK}{KC};$ ∴ $\frac{HB+AH}{HB} = \frac{KC+AK}{KC}.$

∴ $\frac{AB}{HB} = \frac{AC}{KC}.$

∴ $\frac{HB}{AB} = \frac{KC}{AC}.$
Q.E.D.

Also $\frac{HB}{AB} \times \frac{AH}{HB} = \frac{KC}{AC} \times \frac{AK}{KC}.$

∴ $\frac{AH}{AB} = \frac{AK}{AC}.$
Q.E.D.

COROLLARY 2.—If a line HK parallel to BC cuts AB, AC at H, K,

Then $\frac{AH}{AB} = \frac{AK}{AC}$ and $\frac{HB}{AB} = \frac{KC}{AC}.$

COROLLARY 3.—If a line HK cuts AB, AC at H, K so that
$\frac{AH}{AB} = \frac{AK}{AC}$, then HK is parallel to BC.

For riders on Theorems 51, 52 see page 106.

THEOREM 53

If two triangles are equiangular, their corresponding sides are proportional.

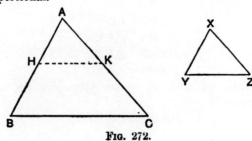

FIG. 272.

PROOFS OF THEOREMS

Given the triangles ABC, XYZ are equiangular, having ∠A = ∠X, ∠B = ∠Y, ∠C = ∠Z.

To Prove $\dfrac{AB}{XY} = \dfrac{AC}{XZ} = \dfrac{BC}{YZ}$.

From AB, AC cut off AH, AK equal to XY, XZ. Join HK.
In the △s AHK, XYZ,

AH = XY, constr.
AK = XZ, constr.
∠HAK = ∠YXZ, given.
∴ △AHK ≡ △XYZ.
∴ ∠AHK = ∠XYZ.

But ∠XYZ = ∠ABC, given.
∴ ∠AHK = ∠ABC.

But these are corresponding angles, ∴ HK is parallel to BC.

$\therefore \dfrac{AB}{AH} = \dfrac{AC}{AK}$.

But AH = XY and AK = XZ.

$\therefore \dfrac{AB}{XY} = \dfrac{AC}{XZ}$.

Similarly it can be proved that $\dfrac{AC}{XZ} = \dfrac{BC}{YZ}$.   Q.E.D.

DEFINITION.—If two polygons are equiangular, and if their corresponding sides are proportional, they are said to be *similar*.

Theorem 53 proves that equiangular triangles are necessarily similar.

THEOREM 54

If the three sides of one triangle are proportional to the three sides of the other, then the triangles are equiangular.

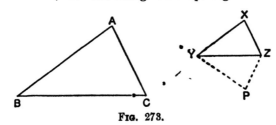

FIG. 273.

Given the △s ABC, XYZ are such that $\dfrac{AB}{XY} = \dfrac{BC}{YZ} = \dfrac{CA}{ZX}$.

To Prove ∠A = ∠X, ∠B = ∠Y, ∠C = ∠Z.

On the side of YZ opposite to X, draw YP and ZP so that ∠ZYP = ∠ABC and ∠YZP = ∠ACB.

Since the △s ABC, PYZ are equiangular, by construction,

$$\dfrac{AB}{YP} = \dfrac{BC}{YZ}.$$

But $\dfrac{AB}{XY} = \dfrac{BC}{YZ}$, given

∴ $\dfrac{AB}{YP} = \dfrac{AB}{XY}.$

∴ YP = XY.

Similarly ZP = XZ.

∴ in the △s XYZ, PYZ.

XY = PY, proved.
XZ = PZ, proved.
YZ is common.

∴ △XYZ ≡ △PYZ.
∴ ∠XYZ = ∠PYZ and ∠XZY = ∠PZY.
But ∠PYZ = ∠ABC and ∠PZY = ∠ACB, constr.
∴ ∠XYZ = ∠ABC and ∠XZY = ∠ACB.
∴ also ∠YXZ = ∠BAC. Q.E.D.

Theorem 55

If two triangles have an angle of one equal to an angle of the other, and the sides about these equal angles proportional, the triangles are equiangular.

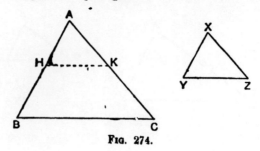

Fig. 274.

PROOFS OF THEOREMS

Given in the triangles ABC, XYZ, $\angle$ BAC $=$ $\angle$ YXZ and $\dfrac{AB}{XY} = \dfrac{AC}{XZ}$.

To Prove $\angle$ ABC $=$ $\angle$ XYZ and $\angle$ ACB $=$ $\angle$ XZY.

From AB, AC, cut off AH, AK equal to XY, XZ. Join HK.

In the $\triangle$s AHK, XYZ,

  AH $=$ XY, constr.
  AK $=$ XZ, constr.
  $\angle$ HAK $=$ $\angle$ YXZ, given.

$\therefore$ $\triangle$ AHK $\equiv$ $\triangle$ XYZ.

$\therefore$ $\angle$ AHK $=$ $\angle$ XYZ and $\angle$ AKH $=$ $\angle$ XZY.

Now $\dfrac{AB}{XY} = \dfrac{AC}{XZ}$ and XY $=$ AH, XZ $=$ AK.

$\therefore$ $\dfrac{AB}{AH} = \dfrac{AC}{AK}$.

$\therefore$ HK is parallel to BC.

$\therefore$ $\angle$ AHK $=$ $\angle$ ABC and $\angle$ AKH $=$ $\angle$ ACB, corresp. $\angle$ s.

But $\angle$ AHK $=$ $\angle$ XYZ and $\angle$ AKH $=$ $\angle$ XZY, proved.

$\therefore$ $\angle$ ABC $=$ $\angle$ XYZ and $\angle$ ACB $=$ $\angle$ XZY.

For riders on Theorems 53, 54, 55 see page 112. Q.E.D.

Theorem 56

(1) If two chords of a circle (produced if necessary) cut one another, the rectangle contained by the segments of the one is equal to the rectangle contained by the segments of the other.

(2) If from any point outside a circle, a secant and a tangent are drawn, the rectangle contained by the whole secant and the part of it outside the circle is equal to the square on the tangent.

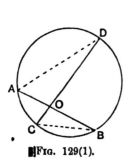
Fig. 129(1).

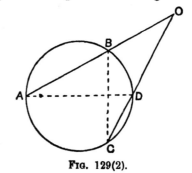
Fig. 129(2).

(1) *Given* two chords **AB, CD** intersecting at **O**.
To Prove **OA . OB = OC . OD**.
Join **BC, AD**.
In the △s **AOD, BOC**,
∠ **OAD** = ∠ **OCB**, in the same segment, Fig. 129(1) and Fig 129(2).
∠ **AOD** = ∠ **COB**, vert. opp. in Fig. 129(1), same ∠ in Fig. 129(2).
∴ the third ∠ **ODA** = the third ∠ **OBC**.
∴ the triangles are equiangular.
∴ $\dfrac{OA}{OC} = \dfrac{OD}{OB}$.
∴ **OA . OB = OC . OD**.   Q.E.D.

(2) *Given* a chord **AB** meeting the tangent at **T** in **O**.
To Prove **OA . OB = OT²**.
Join **AT, BT**.

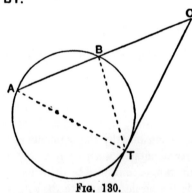

FIG. 130.

In the △s **AOT, TOB**,
∠ **TAO** = ∠ **BTO**, alt. segment.
∠ **AOT** = ∠ **TOB**, same angle.
∴ the third ∠ **ATO** = the third ∠ **TBO**.
∴ the triangles are equiangular.
∴ $\dfrac{OA}{OT} = \dfrac{OT}{OB}$.
∴ **OA . OB = OT²**.   Q.E.D.

Note.—This may also be deduced from (1) by taking the limiting case when **D** coincides with **C** in Fig. 129(2).

PROOFS OF THEOREMS

The converse properties are as follows :—
  (i) If two lines **AOB, COD** are such that **AO . OB = CO . OD**, then **A, B, C, D** lie on a circle.
  (ii) If two lines **OBA, ODC** are such that **OA . OB = OC . OD**, then **A, B, C, D** lie on a circle.
  (iii) If two lines **OBA, OT** are such that **OA . OB = OT²**, then the circle through **A, B, T** touches **OT** at **T**.
These are proved easily by a *reductio ad absurdum* method.

Theorem 57

If **AD** is an altitude of the triangle **ABC**, which is right-angled at **A**, then (i) **AD² = BD . DC**.
    (ii) **BA² = BD . BC**.

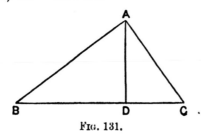

Fig. 131.

(1) Since $\angle$ **BDA** $= 90°$, the remaining angles of the triangle **ABD** add up to $90°$.
$\therefore$ $\angle$ **DAB** $+ \angle$ **DBA** $= 90°$.
But $\angle$ **DAB** $+ \angle$ **DAC** $= 90°$, given.
$\therefore$ $\angle$ **DAB** $+ \angle$ **DBA** $= \angle$ **DAB** $+ \angle$ **DAC**.
$\therefore$ $\angle$ **DBA** $= \angle$ **DAC**.
$\therefore$ in the $\triangle$s **ADB, CDA**,
  $\angle$ **ADB** $= \angle$ **CDA**, right angles.
  $\angle$ **DBA** $= \angle$ **DAC**, proved.
$\therefore$ the third $\angle$ **BAD** $=$ the third $\angle$ **ACD**.
  $\therefore$ the triangles are equiangular.
$\therefore$ $\dfrac{AD}{DC} = \dfrac{BD}{DA}$.
$\therefore$ **AD² = BD . DC**.

Q.E.D.

262   CONCISE GEOMETRY

(2) In the △s **ADB, CAB,**

$\angle$ **ADB** = $\angle$ **CAB**, right angles.
$\angle$ **ABD** = $\angle$ **CBA**, same angle.
∴ the third $\angle$ **DAB** = the third $\angle$ **ACB**.
∴ the triangles are equiangular.

∴ $\dfrac{AB}{BC} = \dfrac{BD}{AB}$

∴ $AB^2 = BD \cdot BC$.   Q.E.D.

An alternative method of proof is given on page 121.

Note.—**AD** is called the *mean proportional* between **BD** and **DC**.
Also **BA** is the *mean proportional* between **BD** and **BC**.

For riders on Theorems 56, 57 see page 122.

Theorem 58

The ratio of the areas of two similar triangles is equal to the ratio of the squares on corresponding sides.

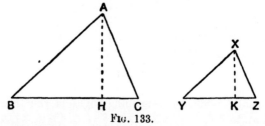

Fig. 133.

Given the triangles **ABC, XYZ** are similar.

To Prove $\dfrac{\triangle ABC}{\triangle XYZ} = \dfrac{BC^2}{YZ^2}$.

Draw the altitudes **AH, XK.**
In the △s **AHB, XKY,**

$\angle$ **ABH** = $\angle$ **XYK**, given.
$\angle$ **AHB** = $\angle$ **XKY**, rt. $\angle$s constr.
∴ the third $\angle$ **BAH** = the third $\angle$ **YXK**.
∴ the △s **AHB, XKY** are similar.

∴ $\dfrac{AH}{XK} = \dfrac{AB}{XY}$.

But $\dfrac{AB}{XY} = \dfrac{BC}{YZ}$, since △s **ABC, XYZ** are similar.

PROOFS OF THEOREMS

$$\therefore \frac{AH}{XK} = \frac{BC}{YZ}.$$

But $\triangle ABC = \tfrac{1}{2} AH \cdot BC$ and $\triangle XYZ = \tfrac{1}{2} XK \cdot YZ$.

$$\therefore \frac{\triangle ABC}{\triangle XYZ} = \frac{AH \cdot BC}{XK \cdot YZ}.$$

But $\dfrac{AH}{XK} = \dfrac{BC}{YZ}$, proved.

$$\therefore \frac{\triangle ABC}{\triangle XYZ} = \frac{BC^2}{YZ^2}.$$

Q.E.D.

If two polygons are similar, it can be proved that they can be divided up into the same number of similar triangles.

Hence it follows that the ratio of the areas of two similar polygons is equal to the ratio of the squares on corresponding sides.

Theorem 59

If three straight lines are proportionals, the ratio of the area of any polygon described on the first to the area of a similar polygon described on the second is equal to the ratio of the first line to the third line.

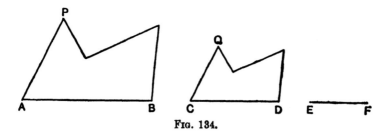

Fig. 134.

Given three lines AB, CD, EF such that $\dfrac{AB}{CD} = \dfrac{CD}{EF}$ and two similar figures ABP, CDQ.

To Prove $\dfrac{\text{figure ABP}}{\text{figure CDQ}} = \dfrac{AB}{EF}.$

Since the figures are similar,

$$\frac{\text{figure ABP}}{\text{figure CDQ}} = \frac{AB^2}{CD^2}.$$

But $\quad CD^2 = AB \cdot EF$, given.

$\therefore \quad \dfrac{AB^2}{CD^2} = \dfrac{AB^2}{AB \cdot EF} = \dfrac{AB}{EF}.$

$\therefore \quad \dfrac{\text{figure ABP}}{\text{figure CDQ}} = \dfrac{AB}{EF}.$

Q.E.D.

For riders on Theorems 58, 59 see page 127.

Theorem 60

(1) If the vertical angle of a triangle is bisected internally or externally by a straight line which cuts the base, or the base produced, it divides the base internally or externally in the ratio of the other sides of the triangle.

(2) If a straight line through the vertex of a triangle divides the base internally or externally in the ratio of the other sides, it bisects the vertical angle internally or externally.

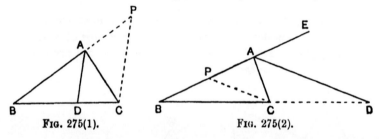

Fig. 275(1).   Fig. 275(2).

(1) *Given* the line **AD** bisecting the angle **BAC**, internally in Fig. 275(1), externally in Fig. 275(2), meets **BC** or **BC** produced at **D**.

To Prove $\dfrac{BD}{DC} = \dfrac{BA}{AC}.$

Through **C** draw **CP** parallel to **DA** to meet **AB** or **AB** produced at **P**.   **BA** is produced to **E** in Fig. 275(2).

In Fig. 275(1). $\angle BAD = \angle APC$, corresp. $\angle$s.
$\angle DAC = \angle ACP$, alt. $\angle$s.
But $\angle BAD = \angle DAC$, given.
$\therefore \quad \angle APC = \angle ACP.$

In Fig. 275(2). $\angle EAD = \angle APC$, corresp. $\angle$s.
$\angle DAC = \angle ACP$, alt. $\angle$s.

PROOFS OF THEOREMS

But $\angle EAD = \angle DAC$, given.

∴ $\angle APC = \angle ACP$.

∴ in each case, $AP = AC$.

But CP is parallel to DA.

∴ $\dfrac{BA}{AP} = \dfrac{BD}{DC}$.

But $AP = AC$, ∴ $\dfrac{BA}{AC} = \dfrac{BD}{DC}$.

Q.E.D.

2) *Given* that AD cuts BC or BC produced so that $\dfrac{BA}{AC} = \dfrac{BD}{DC}$.

To Prove that AD bisects $\angle BAC$ internally or externally.

Through C draw CP parallel to DA to meet AB or AB produced at P.

Now by parallels, $\dfrac{BA}{AP} = \dfrac{BD}{DC}$.

But $\dfrac{BA}{AC} = \dfrac{BD}{DC}$, given.

∴ $\dfrac{BA}{AP} = \dfrac{BA}{AC}$.

∴ $AP = AC$.

∴ $\angle APC = \angle ACP$.

In Fig. 275(1) $\angle APC = \angle BAD$, corresp. $\angle$s.

$\angle ACP = \angle DAC$, alt. $\angle$s.

But $\angle APC = \angle ACP$, proved.

∴ $\angle BAD = \angle DAC$.

In Fig. 275(2) $\angle APC = \angle EAD$, corresp. $\angle$s.

$\angle ACP = \angle DAC$, alt. $\angle$s.

But $\angle APC = \angle ACP$, proved.

∴ $\angle EAD = \angle DAC$.

∴ in Fig. 275(1), AD bisects $\angle BAC$ internally, and in Fig. 275(2), AD bisects $\angle BAC$ externally

Q.E.D.

For riders on Theorem 60 see page 132.

CONSTRUCTIONS FOR BOOK I

CONSTRUCTION 1

From a given point in a given straight line, draw a straight line making with the given line an angle equal to a given angle.

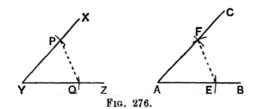

FIG. 276.

Given a point **A** on a given line **AB** and an angle **XYZ**.
   To Construct a line **AC** such that $\angle\,\textbf{CAB} = \angle\,\textbf{XYZ}$.
   With centre **Y** and any radius, draw an arc of a circle cutting **YX**, **YZ** at **P, Q**.
   With centre **A** and the same radius, draw an arc of a circle **EF**, cutting **AB** at **E**.
   With centre **E** and radius equal to **QP**, describe an arc of a circle, cutting the arc **EF** at **F**.
   Join **AF** and produce it to **C**.
   Then **AC** is the required line.
   Proof. Join **PQ, EF**.
   In the △s **PYQ, FAE**,
        YP = **AF**, constr.
        YQ = **AE**, constr.
        PQ = **EF**, constr.
  ∴ △**PYQ** ≡ △**FAE**.
  ∴ $\angle\,\textbf{XYZ} = \angle\,\textbf{CAB}$.

                                          Q.E.F.

Construction 2

Bisect a given angle.

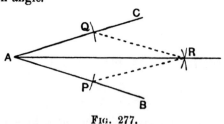

Fig. 277.

Given an angle **BAC**.

To Construct a line bisecting the angle.

With **A** as centre and any radius, draw an arc of a circle, cutting **AB**, **AC** at **P**, **Q**.

With centres **P**, **Q** and with any sufficient radius, the same for each, draw arcs of circles, cutting at **R**. Join **AR**.

Then **AR** is the required bisector.

Proof. Join **PR**, **QR**.

In the △s **APR, AQR**,

   AP = **AQ**, radii of the same circle.

   PR = **QR**, radii of equal circles.

   AR is common.

∴ △**APR** ≡ △**AQR**.

∴ ∠**PAR** = ∠**QAR**.

Q.E.F.

Construction 3

Draw the perpendicular bisector of a given finite straight line.

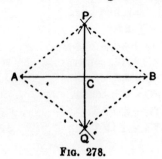

Fig. 278.

CONSTRUCTIONS FOR BOOK I 269

Given a finite line **AB**.
  To Construct the line bisecting **AB** at right angles.
  With centres **A, B** and any sufficient radius, the same for each, draw arcs of circles to cut at **P, Q**.
  Join **PQ** and let it cut **AB** at **C**.
  Then **C** is the mid-point of **AB**, and **PCQ** bisects **AB** at right angles.
  Proof. Join **PA, PB, QA, QB**.
  In the △s **PAQ, PBQ**,
          PA = PB, radii of equal circles.
          QA = QB, radii of equal circles.
          PQ is common.
  ∴ △**PAQ** ≡ △**PBQ**.
  ∴ ∠**APQ** = ∠**BPQ**.
  In the △s **APC, BPC**,
          PA = PB, radii of equal circles.
          PC is common.
      ∠**APC** = ∠**BPC**, proved.
  ∴ △**APC** ≡ △**BPC**.
  ∴   **AC = CB**.
  and ∠**ACP** = ∠**BCP**.
  But these are adjacent angles, ∴ each is a right angle.

Q.E.F.

Construction 4

Draw a straight line at right angles to a given straight line from a given point in it.

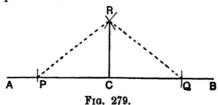

Fig. 279.

Given a point **C** on a line **AB**.
  To Construct a line from **C** perpendicular to **AB**.
  With centre **C** and any radius, draw an arc of a circle cutting **AB** at **P, Q**.

With centres **P, Q** and any sufficient radius, the same for each, draw arcs of circles to cut at **R**. Join **CR**.

Then **CR** is the required perpendicular.

Proof. Join **PR, QR**.

In the △s **RCP, RCQ**,

   RP = RQ, radii of equal circles.

   CP = CQ, radii of the same circle.

   CR is common.

∴ △**RCP** ≡ △**RCQ**.

∴ ∠ **RCP** = ∠ **RCQ**.

But these are adjacent angles, ∴ each is a right angle.

Q.E.F.

Construction 5

Draw a perpendicular to a given straight line of unlimited length from a given point outside it.

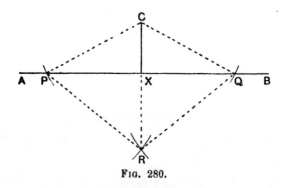

Fig. 280.

Given a line **AB** and a point **C** outside it.

To Construct a line from **C** perpendicular to **AB**.

With **C** as centre and any sufficient radius, draw an arc of a circle, cutting **AB** at **P, Q**.

With **P, Q** as centres and any sufficient radius, the same for each, draw arcs of circles, cutting at **R**. Join **CR** and let it cut **AB** at **X**.

Then **CX** is perpendicular to **AB**.

Proof. Join **CP, CQ, RP, RQ**.

In the △s CPR, CQR,
 CP = CQ, radii of the same circle.
 RP = RQ, radii of equal circles.
 CR is common.
∴ △CPR ≡ △CQR.
∴ ∠PCR = ∠QCR.
In the △s CPX, CQX,
 CP = CQ, radii.
 CX is common.
 ∠PCX = ∠QCX, proved.
∴ △CPX ≡ △CQX.
∴ ∠CXP = ∠CXQ.
But these are adjacent angles, ∴ each is a right angle.

Q.E.F.

Construction 6

Through a given point, draw a straight line parallel to a given straight line.

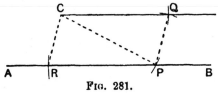

Fig. 281.

Given a line AB and a point C outside it.
To Construct a line through C parallel to AB.
With C as centre and any sufficient radius, draw an arc of a circle PQ, cutting AB at P.
With P as centre and the same radius, draw an arc of a circle, cutting AB at R.
With centre P and radius equal to CR, draw an arc of a circle, cutting the arc PQ at Q on the same side of AB as C. Join CQ.
Then CQ is parallel to AB.
Proof. Join CR, CP, PQ.
In the △s CRP, PQC,
 CR = PQ, constr.
 RP = QC radii of equal circles.
 PC is common.

∴ △CRP ≡ △PQC.
∴ ∠CPR = ∠PCQ.

But these are alternate angles, ∴ CQ is parallel to RP.

Q.E.F.

Construction 7

Draw a triangle having its sides equal to three given straight lines, any two of which are together greater than the third side.

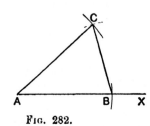

Fig. 282.

Given three lines a, b, c.

To Construct a triangle whose sides are respectively equal to a, b, c.

Take any line **AX**, and with **A** as centre and radius equal to c, draw an arc of a circle, cutting **AX** at **B**.

With **A** as centre and radius equal to b, draw an arc of a circle; and with **B** as centre and radius equal to a, draw an arc of a circle, cutting the former arc at **C**.

Join **AC**, **BC**.

Then **ABC** is the required triangle.

Proof. By construction, **AB** = c.
AC = b.
BC = a.

Q.E.F.

Construction 8

Draw a triangle, given two angles and the perimeter.

Given two angles **X**, **Y** and a line **HK**.

To Construct a triangle having two of its angles equal to **X** and **Y** and its perimeter equal to **HK**.

CONSTRUCTIONS FOR BOOK I 273

Construct lines PH, QK on the same side of HK such that ∠PHK = ∠X and ∠QKH = ∠Y.

Construct lines HA, KA intersecting at A and bisecting the angles PHK, QKH.

Construct through A, lines AB, AC parallel to PH, QK, cutting HK at B, C.

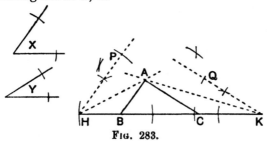

Fig. 283.

Then ABC is the required triangle.

Proof.   ∠BAH = ∠AHP, since AB is parallel to PH.
            ∠BHA = ∠AHP, constr.
∴   ∠BAH = ∠BHA.
∴   BH = BA.

Similarly it may be proved that CK = CA.

∴   AB + BC + CA = HB + BC + CK = HK.

Also ∠ABC = ∠PHK = ∠X, corresp. ∠s.
and ∠ACB = ∠QKH = ∠Y, corresp. ∠s.

∴ ABC is the required triangle.

Q.E.F.

CONSTRUCTION 9

Draw a triangle given one angle, the side opposite that angle and the sum of the other two sides.

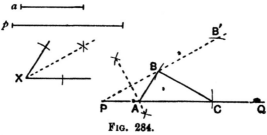

Fig. 284.

Given two lines a, p and an angle **X**.

To Construct a triangle **ABC** such that $BC = a$, $BA + AC = p$, $\angle BAC = \angle X$.

Draw a line **PQ** and from it cut off a part **PC** equal to p. Construct a line **PB** such that $\angle BPC$ equals $\frac{1}{2} \angle X$. With **C** as centre and radius equal to a, draw an arc of a circle, cutting **PB** at **B**. Construct the perpendicular bisector of **PB** and let it meet **PC** at **A**. Join **AB**, **BC**.

Then **ABC** is the required triangle

Proof. Since **A** lies on the perpendicular bisector of **PB**,
$$AP = AB.$$
$\therefore \quad \angle APB = \angle ABP.$
$\therefore \quad \angle BAC = \angle APB + \angle ABP.$
$\qquad\qquad = 2 \angle APB.$
$\qquad\qquad = \angle X$ since $\angle APB = \frac{1}{2} \angle X.$

Also $AB + AC = AP + AC = PC = p$.
and $BC = a$, by construction.
$\therefore$ **ABC** is the required triangle.

Q.E.F.

Note.—Since there are two possible positions of **B**, namely, **B** and **B′**, there are two triangles which satisfy the given conditions.

Construction 10

Given the angle **BAC**, construct points **P**, **Q** on **AB**, **AC** such that **PQ** is of given length and the angle **APQ** of given size.

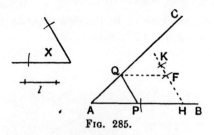

Fig. 285.

Given the angle **BAC**, a line l and an angle **X**.

To Construct points **P**, **Q** on **AB**, **AC** such that **PQ** equals l and $\angle APQ$ equals $\angle X$.

CONSTRUCTIONS FOR BOOK I 275

Take any point H on AB and construct a line HK such that
∠AHK = ∠X.
From HK cut off HF equal to l. Through F draw FQ parallel to AB to cut AC in Q. Through Q draw QP parallel to FH to cut AB in P.
Then PQ is the required line.
Proof. By construction, PQFH is a parallelogram,
∴ PQ = HF = l;
and ∠QPA = ∠FHA = ∠X,
∴ PQ is the required line.

Q.E.F.

Construction 11

Describe a square on a given straight line.

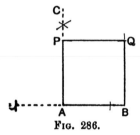

Fig. 286.

Given a line AB.
To Construct a square on AB.
From A draw a line AC perpendicular to AB; from AC cut off AP equal to AB.
Through P draw PQ parallel to AB.
Through B draw BQ parallel to AP, cutting PQ at Q.
Then ABQP is the required square.
Proof. By construction, ABQP is a parallelogram.
But ∠BAP = 90°, ∴ ABQP is a rectangle.
But AB = AP, ∴ ABQP is a square.

Q.E.F.

CONSTRUCTIONS FOR BOOK II

Construction 12

(1) Reduce a quadrilateral to a triangle of equal area.
(2) Reduce any given rectilineal figure to a triangle of equal area.

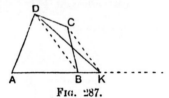

Fig. 287.

(1) *Given* a quadrilateral **ABCD**.
  To Construct a triangle equal in area to it.
  Join **BD**.
  Through **C**, draw **CK** parallel to **DB** to meet **AB** produced at **K**. Join **DK**.
  Then **ADK** is the required triangle.
  Proof. The triangles **BCD**, **BKD** are on the same base **BD** and between the same parallels **BD**, **KC**.
  ∴ area of △**BCD** = area of △**BKD**.
  Add to each △**ABD**.
  ∴ area of quad. **ABCD** = area of △**AKD**.
  ∴ **AKD** is the required triangle.   Q.E.F.

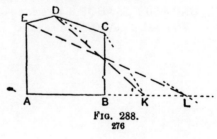

Fig. 288.

CONSTRUCTIONS FOR BOOK II   277

(2) *Given* a pentagon **ABCDE**.
   To Construct a triangle equal in area to it.
   Proceed as in (1). This reduces the pentagon to a quadrilateral **AKDE** of equal area.
   Proceed as in (1). Join **EK**; through **D** draw **DL** parallel to **EK** to meet **AK** produced in **L**.
   Then **AEL** is the required triangle.
   This process can be repeated any number of times.

Construction 13

Bisect a triangle by a line through a given point in one side.

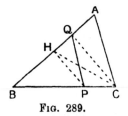

Fig. 289.

Given a point **P** on the side **BC** of the triangle **ABC**.
   To Construct a line **PQ** bisecting the triangle.
   Suppose **P** is nearer to **C** than **B**.
   Bisect **AB** at **H**. Join **PH**.
   Through **C**, draw **CQ** parallel to **PH** to meet **AB** at **Q**.
   Join **PQ**.
   Then **PQ** is the required line.
   Proof. Join **CH**.
   Since **AH = HB**, area of $\triangle$**AHC** = area of $\triangle$**BHC**.
   $\therefore$ $\triangle$**BHC** = $\frac{1}{2}$$\triangle$**ABC**.
   Since **HP** is parallel to **QC**.
   Area of $\triangle$**HPQ** = area of $\triangle$**HPC**.
   Add to each, $\triangle$**BHP**.
   $\therefore$ area of $\triangle$**BPQ** = area of $\triangle$**BHC**.
   But $\triangle$**BHC** = $\frac{1}{2}$$\triangle$**ABC**.
   $\therefore$ $\triangle$**BPQ** = $\frac{1}{2}$$\triangle$**ABC**.
   $\therefore$ **PQ** bisects $\triangle$**ABC**.

Q.E.F.

If it is required to draw a line **PQ**, cutting off from the triangle **ABC** a triangle **BPQ** equal to a given fraction, say $\frac{2}{7}$, of the triangle **ABC**, take a point **H** on **BA** such that **BH** $=\frac{2}{7}$**BA** and proceed as in the above Construction.

Construction 14

Divide a given straight line into any given number of equal parts.

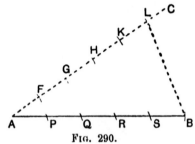

Fig. 290.

Given a line **AB**.

To Construct points dividing **AB** into any number (say 5) equal parts.

Through **A**, draw any line **AC**.

Along **AC**, step out with compasses equal lengths, the number of such lengths being the required number of equal parts (in this case 5).

Let the equal lengths be **AF, FG, GH, HK, KL**.

Join **LB**, and through **F, G, H, K** draw lines parallel to **BL**, meeting **AB** at **P, Q, R, S**.

Then **AP, PQ, QR, RS, SB** are the required equal parts.

Proof. Since the parallel lines **FP, GQ, HR, KS, LB** cut off equal intercepts on **AC**, they cut off equal intercepts on **AB**.

Q.E.F.

Construction 15

Divide a quadrilateral into any number of equal parts by lines through one vertex.

Given a quadrilateral **ABCD**.

To Construct lines through **A** which divide **ABCD** into any number (say 5) equal parts.

Join **AC**; through **D** draw **DP** parallel to **AC** to meet **BC** produced at **P**.

CONSTRUCTIONS FOR BOOK II 279

Divide **BP** into the required number (in this case 5) of equal parts, BQ_1, Q_1Q_2, Q_2Q_3, Q_3Q_4, Q_4P.

Through those points which lie on **BC** produced, in this case Q_3Q_4, draw lines Q_3R_3, Q_4R_4 parallel to **PD** to meet **CD** in R_3, R_4.

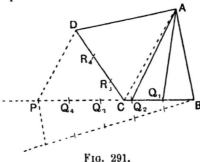

Fig. 291.

Then AQ_1, AQ_2, AR_3, AR_4 are the required lines.

Proof. By construction, the $\triangle ABP$ and the quad. **ABCD** are equal in area.

But the areas of $\triangle$s BAQ_1, Q_1AQ_2, Q_2AQ_3, Q_3AQ_4, Q_4AP are equal, for their bases are equal and they have the same height.

∴ each $= \frac{1}{5} \triangle ABP = \frac{1}{5}$ quad. **ABCD**.

Further, $\triangle ACQ_3 = \triangle ACR_3$, $\triangle ACQ_4 = \triangle ACR_4$, $\triangle ACP = \triangle ACD$, being on the same base and between the same parallels.

∴ $\triangle AR_3R_4 = \triangle ACR_4 - \triangle ACR_3 = \triangle ACQ_4 - \triangle ACQ_3 = \triangle AQ_3Q_4$.

And similarly $\triangle AR_4D = \triangle AQ_4P$.

Also quad. $AQ_2CR_3 = \triangle AQ_2C + \triangle ACR_3$.
$\qquad = \triangle AQ_2C + \triangle ACQ_3$.
$\qquad = \triangle AQ_2Q_3$.

∴ AQ_1, AQ_2, AR_3, AR_4 divide quad. **ABCD** into five equal parts. Q.E.F.

Note.—The same method may be used for dividing a rectilineal figure with any number of sides into any number of equal parts, either by lines through a vertex or by lines through a given point on one of the sides.

CONSTRUCTIONS FOR BOOK III

CONSTRUCTION 16

Construct the centre of a circle, an arc of which is given.

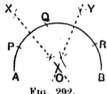

Fig. 292.

Given an arc AB of a circle.

To Construct the centre of the circle.

Take three points P, Q, R on the arc.
Construct the perpendicular bisectors OX, OY of PQ, QR, intersecting at O.
Then O is the required centre.

Proof. The perpendicular bisector of a chord of a circle passes through the centre of the circle.
∴ the centre of the circle lies on OX and on OY.
∴ the centre is at O.

Q.E.F.

CONSTRUCTION 17

Construct a circle to pass through three given points, which do not lie on a straight line.

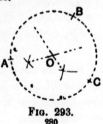

Fig. 293.

CONSTRUCTIONS FOR BOOK III   281

Given three points A, B, C.
  To Construct a circle to pass through **A, B, C**.
  Construct the perpendicular bisectors **OX, OY** of **AB, BC**, intersecting at **O**.
  With **O** as centre and **OA** as radius, describe a circle.
  This is the required circle.
  Proof. Since **O** lies on the perp. bisector of **AB**,
$$OA = OB.$$
Since **O** lies on the perp. bisector of **BC**,
$$OB = OC.$$
$$\therefore\ OA = OB = OC.$$
∴ the circle, centre **O**, radius **OA**, passes through **B, C**.
                                                  Q.E.F.

Construction 18

(1) Construct the inscribed circle of a given triangle.
(2) Construct an escribed circle of a given triangle.

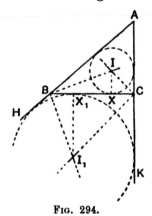

Fig. 294.

Given a triangle **ABC**.
  To Construct (1) the circle inscribed in △**ABC**.
                 (2) the circle which touches **AB** produced, **AC** produced and **BC**.
(1) Construct the lines **BI, CI**, bisecting the angles **ABC, ACB** and intersecting at **I**.
  Draw **IX** perpendicular to **BC**.
  With **I** as centre and **IX** as radius, describe a circle.

282 CONCISE GEOMETRY

This circle touches **BC, CA, AB**.

Proof. Since **I** lies on the bisector of ∠ **ABC**,
I is equidistant from the lines **BA, BC**.
Since **I** lies on the bisector of ∠ **ACB**.
I is equidistant from the lines **CB, CA**.
∴ **I** is equidistant from **AB, BC, CA**.
∴ the circle, centre **I**, radius **IX**, touches **AB, BC, CA**.

(2) Produce **AB, AC** to **H, K**. Construct the lines **BI$_1$, CI$_1$**, bisecting the angles **HBC, KCB** and intersecting at **I$_1$**.
Draw **I$_1$X$_1$** perpendicular to **BC**.
With **I$_1$** as centre and **I$_1$X$_1$** as radius, describe a circle.
This circle touches **AB** produced, **AC** produced and **BC**.

Proof. Since **I$_1$** lies on the bisector of ∠ **HBC**,
I$_1$ is equidistant from **BH** and **BC**.
Since **I$_1$** lies on the bisector of ∠ **KCB**,
I$_1$ is equidistant from **CK** and **CB**
∴ **I$_1$** is equidistant from **HB, BC, CK**.
∴ the circle, centre **I$_1$**, radius **I$_1$X$_1$**, touches **HB, BC, CK**.

Q.E.F.

CONSTRUCTION 19

(1) Construct a tangent to a circle at a given point on the circumference.
(2) Construct the tangents to a circle from a given point outside it

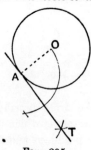

Fig. 295.

(1) *Given* a point **A** on the circumference of a circle.
To Construct the tangent at **A** to the circle.
Construct the centre **O** of the circle. Join **AO**.

CONSTRUCTIONS FOR BOOK III 283

Through **A**, construct a line **AT** perpendicular to **AO**.
Then **AT** is the required tangent.
Proof. The tangent is perp. to the radius through the point of contact. But **AO** is a radius and $\angle$ **OAT** $= 90°$,
∴ **AT** is the tangent at **A**.

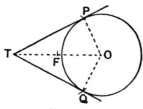

Fig. 296.

(2) *Given* a point **T** outside a circle.
To Construct the tangents from **T** to the circle.
Construct the centre **O** of the circle. Join **OT** and bisect it at **F**. With centre **F** and radius **FT**, describe a circle and let it cut the given circle at **P, Q**. Join **TP, TQ**.
Then **TP, TQ** are the required tangents.
Proof. Since **TF** = **FO**, the circle, centre **F**, radius **FT**, passes through **O**, and **TO** is a diameter.
∴ $\angle$ **TPO** $= 90° =$ $\angle$ **TQO**. $\angle$ in semicircle.
But **OP, OQ** are radii of the given circle.
∴ **TP, TQ** are tangents to the given circle.

Q.E.F.

CONSTRUCTION 20

(1) Draw the direct (or exterior) common tangents to two circles.
(2) Draw the transverse (or interior) common tangents to two non-intersecting circles.

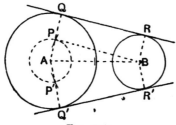

Fig. 297.

(1) *Given* two circles, centres **A, B**.

To Construct their direct common tangents.

Let a, b be the radii of the circles, centres **A, B**, and suppose $a > b$. With **A** as centre and $a - b$ as radius, describe a circle and construct the tangents **BP, BP′** from **B** to this circle. Join **AP, AP′** and produce them to meet the circle, radius a, in **Q, Q′**. Through **Q, Q′** draw lines **QR, Q′R′** parallel to **PB, P′B**.

Then **QR, Q′R′** are the required common tangents.

Proof. Draw **BR, BR′** parallel to **AQ, AQ′** to meet **QR, Q′R′** at **R, R′**.

By Construction, **PQRB** is a parallelogram.

∴ **BR** = **PQ** = **AQ** − **AP** = $a - (a - b) = b$.

∴ **R** lies on the circle, centre **B**, radius b.

Also, since **BP** is a tangent, ∠**BPA** = 90°.

∴ ∠**RQA** = 90° and ∠**BRQ** = 90°, by parallels.

∴ **QR** is a tangent at **Q** and **R** to the two circles.

Similarly it may be proved that **Q′R′** is also a common tangent.

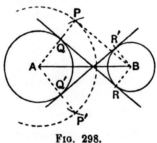

Fig. 298.

(2) *Given.* Two non-intersecting circles, centres **A, B**.

To Construct the transverse common tangents.

Let a, b be the radii of the circles, centres **A, B**.

With **A** as centre and $a + b$ as radius, describe a circle and construct the tangents **BP, BP′** to it from **B**.

Join **AP, AP′**, cutting the circle radius a at **Q, Q′**.

Through **Q, Q′** draw lines **QR, Q′R′** parallel to **PB, P′B**.

Then **QR, Q′R′** are the required common tangents.

CONSTRUCTIONS FOR BOOK III   285

Proof. Through B draw BR, BR' parallel to AQ, AQ' to meet QR, Q'R' at R, R'.
By construction, PBRQ is a parallelogram.
∴ BR = PQ = AP - AQ = $(a+b) - a = b$.
∴ R lies on the circle, centre B, radius b.
Also, since BP is a tangent, ∠BPA = 90°.
∴ ∠AQR = 90° and ∠BRQ = 90°, by parallels.
∴ QR is a tangent at Q and R to the two circles.
Similarly it may be proved that Q' R' is also a common tangent.

Q.E.F.

Construction 21

On a given straight line, construct a segment of a circle containing an angle equal to a given angle.

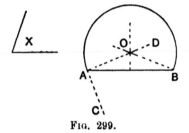

Fig. 299.

Given a straight line AB and an angle X.
To Construct on AB a segment of a circle containing an angle equal to ∠X.
At A, make an angle BAC equal to ∠X.
Draw AD perpendicular to AC.
Draw the perpendicular bisector of AB and let it meet AD at O.
With O as centre and OA as radius, describe a circle.
Then the segment of this circle on the side of AB opposite to C is the required segment.
Proof. Since O lies on the perpendicular bisector of AB, OA = OB; ∴ the circle passes through B.
Since AC is perpendicular to the radius OA, AC is a tangent;
∴ ∠X = ∠CAB = angle in alternate segment.

Q.E.F.

Construction 22

Inscribe in a given circle a triangle equiangular to a given triangle.

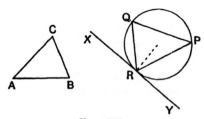

Fig. 300.

Given a circle and a triangle ABC.

To Construct a triangle inscribed in the circle and equiangular to ABC.

Take any point R on the circle and construct the tangent XRY at R to the circle.

Draw chords RP, RQ so that $\angle$ PRY = $\angle$ CBA and $\angle$ QRX = $\angle$ CAB.

Join PQ.

Then PQR is the required triangle.

Proof. $\angle$ PQR = $\angle$ PRY, alt. segment.
        = $\angle$ CBA.
and $\angle$ QPR = $\angle$ QRX, alt. segment.
        = $\angle$ CAB.
$\therefore$ the remaining $\angle$ QRP = the remaining $\angle$ BCA.

Q.E.F.

Construction 23

Describe about a given circle a triangle equiangular to a given triangle.

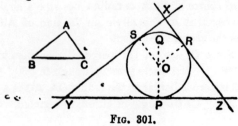

Fig. 301.

CONSTRUCTIONS FOR BOOK III 287

Given a circle and a triangle **ABC**.

To Construct a triangle with its sides touching the circle and equiangular to **ABC**.

Construct the centre **O** of the circle: draw any radius **OP** and produce **PO** to **Q**. Draw radii **OR**, **OS** so that $\angle$ **QOR** = $\angle$ **ACB** and $\angle$ **QOS** = $\angle$ **ABC**. Draw the tangents at **P**, **R**, **S**, forming the triangle **XYZ**.

Then **XYZ** is the required triangle.

Proof. $\angle$ **ORZ** = $90°$ = $\angle$ **OPZ** since **PZ**, **RZ** are tangents.

∴ **ORZP** is a cyclic quadrilateral.

∴ $\angle$ **QOR** = $\angle$ **PZR**, ext. $\angle$ cyclic quad. = int. opp. $\angle$.

But $\angle$ **QOR** = $\angle$ **ACB**, constr.

∴ $\angle$ **PZR** = $\angle$ **ACB**.

Similarly $\angle$ **PYS** = $\angle$ **ABC**.

∴ the remaining $\angle$ **YXZ** of the $\triangle$ **XYZ** = the remaining $\angle$ **BAC**.

Q.E.F.

CONSTRUCTION 24

Construct a circle to pass through a given point **A** and to touch a given circle at a given point **B**.

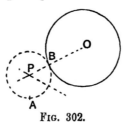

FIG. 302.

Construct the centre **O** of the given circle.

Construct the perpendicular bisector of **AB** and produce it to cut **OB**, or **OB** produced at **P**.

With **P** as centre and **PB** as radius, describe a circle. This is the required circle.

Proof. Since **P** lies on the perpendicular bisector of **AB**,
  PA = **PB**.

Since **P** lies on **OB**, or **OB** produced, the two circles touch at **B**.

Q.E.F.

288   CONCISE GEOMETRY

Construction 25

Construct a circle to touch a given circle and to touch a given line **ABC** at a given point **B** on it.

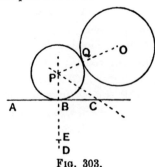

Fig. 303.

Construct the centre **O** of the given circle.
Construct the perpendicular **BD** to **AB** and cut off a part **BE** equal to the radius of the given circle.
Construct the perpendicular bisector of **OE** and produce it to cut **EB**, or **EB** produced at **P**.
With **P** as centre and **PB** as radius, describe a circle.
This is the required circle.
[There are two solutions according to which side **E** is of **AC**.]
Proof. Let **PO** cut the given circle at **Q**.
P lies on the perpendicular bisector of **OE**.
∴ **PO = PE**.
In Fig. 303, **BE = OQ** and **PE = PO**.
∴ **PB = PQ**.
Also **P** lies on **OQ** produced.
∴ the circle, centre **P**, radius **PB**, passes through **Q** and touches the given circle at **Q**.

Q.E.F.

CONSTRUCTIONS FOR BOOK IV

CONSTRUCTION 26

Divide a given finite straight line in a given ratio (i) internally, (ii) externally.

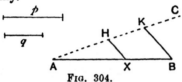

FIG. 304.

Given two lines p, q and a finite line AB.

To Construct (i) a point X in AB such that $\dfrac{AX}{XB} = \dfrac{p}{q}$.

(ii) a point Y in AB produced such that $\dfrac{AY}{BY} = \dfrac{p}{q}$.

(i) Draw any line AC and cut off successively AH = p, HK = q. Join KB. Through H draw a line parallel to KB to cut AB at X.

Then $\dfrac{AX}{XB} = \dfrac{AH}{HK} = \dfrac{p}{q}$ by parallels. Q.E.F.

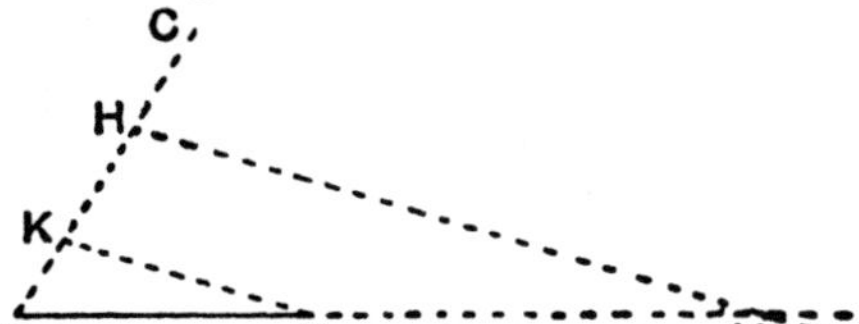

FIG. 305.

(ii) Draw any line AC; cut off AH = p, and from HA cut off HK = q. Join KB. Through H draw a line parallel to KB to cut AB produced at Y.

Then $\dfrac{AY}{BY} = \dfrac{AH}{KH} = \dfrac{p}{q}$ by parallels. Q.E.F.

CONSTRUCTION 27

Construct a fourth proportional to three given lines.

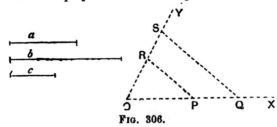

Fig. 306.

Given three lines of lengths a, b, c units.

To Construct a line of length d units, such that $\dfrac{a}{b} = \dfrac{c}{d}$.

Draw any two lines **OX**, **OY**.
From **OX** cut off parts **OP**, **OQ** such that **OP** = a, **OQ** = b.
From **OY** cut off a part **OR** such that **OR** = c.
Join **PR**.
Through **Q**, draw a line **QS** parallel to **PR** to meet **OY** at **S**.
Then **OS** is the required fourth proportional.

Proof. Since **PR** is parallel to **QS**

$$\frac{OP}{OQ} = \frac{OR}{OS},$$

$$\therefore \frac{a}{b} = \frac{c}{OS}.$$

Q.E.F.

Note.—To construct a third proportional to two given lines, lengths a, b units, is the same as constructing a fourth proportional to three lines of length a, b, b units.

CONSTRUCTION 28

To construct a polygon similar to a given polygon and such that corresponding sides are in a given ratio.

Given a polygon **OABCD** and a ratio **XY** : **XZ**.

To Construct a polygon **OA'B'C'D'** such that $\dfrac{OA'}{OA} = \dfrac{A'B'}{AB} = \ldots = \dfrac{XY}{XZ}.$

CONSTRUCTIONS FOR BOOK IV 291

Join **OB, OC**.

Draw any line **OQ** and cut off parts **OP′, OP** equal to **XY, XZ**. Join **PA**.

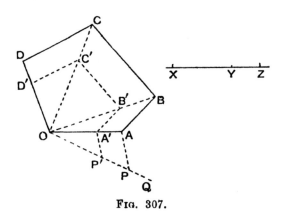

FIG. 307.

Through **P′** draw **P′A′** parallel to **PA** to meet **OA** at **A′**.
Through **A′** draw **A′B′** parallel to **AB** to meet **OB** at **B′**.
Through **B′** draw **B′C′** parallel to **BC** to meet **OC** at **C′**.
Through **C′** draw **C′D′** parallel to **CD** to meet **OD** at **D′**.
Then **OA′B′C′D′** is the required polygon.

Proof. Since **A′B′** is parallel to **AB**, △s **OA′B′, OAB** are similar,

$$\therefore \frac{OA'}{OA} = \frac{A'B'}{AB} = \frac{OB'}{OB}.$$

Similarly $\frac{OB'}{OB} = \frac{B'C'}{BC} = \frac{OC'}{OC}$, and so on.

$$\therefore \frac{OA'}{OA} = \frac{A'B'}{AB} = \frac{B'C'}{BC} = \frac{C'D'}{CD} = \frac{D'O}{DO}.$$

Also $\frac{OA'}{OA} = \frac{OP'}{OP} = \frac{XY}{XZ}.$

∴ the sides of **OA′B′C′D′** are proportional to the sides of **OABCD** in the ratio **XY : XZ**.

Further, by parallels, the polygons are equiangular.

∴ the polygons are similar and their corresponding sides are in the given ratio.

Q.E.F.

292 CONCISE GEOMETRY

CONSTRUCTION 29

Inscribe a square in a given triangle.

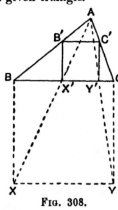

Fig. 308.

Given a triangle **ABC**.

To Construct a square with one side on **BC** and its other corners on **AB** and **AC**.

On **BC** describe the square **BXYC**.
Join **AX, AY**, cutting **BC** at **X′, Y′**.
Through **X′, Y′** draw **X′B′, Y′C′** parallel to **XB** (or **YC**) to cut **AB, AC** at **B′, C′**. Join **B′C′**.
Then **B′X′Y′C′** is the required square.

Proof. By parallels $\dfrac{AB'}{AB} = \dfrac{B'X'}{BX} = \dfrac{AX'}{AX} = \dfrac{X'Y'}{XY} = \dfrac{AY'}{AY} = \dfrac{Y'C'}{YC} = \dfrac{AC'}{AC}$.

∴ Since $\dfrac{AB'}{AB} = \dfrac{AC'}{AC}$, **B′C′** is parallel to **BC** and $\dfrac{B'C'}{BC} = \dfrac{AB'}{AB}$.

∴ **B′X′Y′C′** is similar to **BXYC** and is ∴ a square.

Q.E.F.

The following is a more general but less neat method.

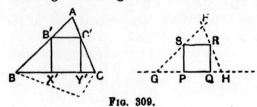

Fig. 309.

Take any square **PQRS** with **PQ** parallel to **BC**, and circumscribe a triangle **FGH** about this square equiangular to **ABC**. [Draw **SF**, **RF** parallel to **AB**, **AC**; produce **FS**, **FR** to meet **PQ** produced at **G**, **H**.]

Divide **BC** at **X'** in the ratio **GP** : **PH**.

Then **X'** is one corner of the square; complete by parallels and perpendiculars.

Construction 30

Construct a mean proportional to two given lines.

Given two lines of lengths a, b units.

To Construct a line of length x units such that $\dfrac{a}{x} = \dfrac{x}{b}$ or $x^2 = ab$

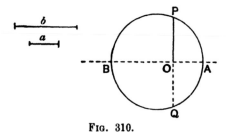

Fig. 310.

METHOD I.— Take a point **O** on a line and cut off from the line on *opposite* sides of **O**, parts **OA**, **OB** of lengths a, b units.

On **AB** as diameter, describe a circle.

Draw **OP** perpendicular to **AB** to cut the circle at **P**.

Then **OP** is the required mean proportional.

Proof. Produce **PO** to meet the circle at **Q**.

PQ is a chord perpendicular to the diameter **AB**,

$\therefore$ **PO** = **OQ**.

But **PO** . **OQ** = **AO** . **OB**, intersecting chords of a circle.

$\therefore$ $OP^2 = a \cdot b$

or $\dfrac{a}{OP} = \dfrac{OP}{b}.$

Q.E.F.

METHOD II.—Take a point **O** on a line and cut off from the line on the *same* side of **O**, parts **OA**, **OB** of lengths a, b units.

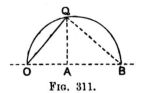

FIG. 311.

On **OB** as diameter, describe a circle.
Draw **AQ** perpendicular to **OB** to meet the circle at **Q**.
Join **OQ**. Then **OQ** is the required mean proportional.
Proof. $\angle \mathbf{OQB} = 90°$; angle in semicircle.
∴ **OQ** is a tangent to the circle on **QB** as diameter.
But $\angle \mathbf{QAB} = 90°$, ∴ circle on **QB** as diameter passes through **A**.
∴ $\mathbf{OQ}^2 = \mathbf{OA} \cdot \mathbf{OB}$, tangent property of circle.
∴ $\mathbf{OQ}^2 = a \cdot b$ or $\dfrac{a}{\mathbf{OQ}} = \dfrac{\mathbf{OQ}}{b}$.

Q.E.F.

Note.—In practical constructions, Method II. is often preferable to Method I.

CONSTRUCTION 31

(i) Construct a square equal in area to a given rectangle.
(ii) Construct a square equal in area to a given polygon.

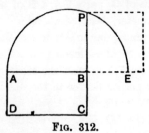

FIG. 312.

(i) *Given* a rectangle **ABCD**.
  To Construct a square of equal area.
  Produce **AB** to **E**, making **BE** = **BC**.

On **AE** as diameter, describe a semicircle.
Produce **CB** to meet the semicircle at **P**.
On **BP** describe a square.
This is the required square.

Proof. By the proof of Constr. 30, $BP^2 = AB \cdot BE$, but
$BE = BC$.
∴ $BP^2 = AB \cdot BC$ = area of **ABCD**.

Q.E.F.

(ii) *Given* any polygon.
To Construct a square of equal area.

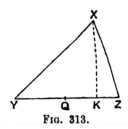

Fig. 313.

By the method of Constr. 12, reduce the polygon to an equivalent triangle **XYZ**.
Draw the altitude **XK** and bisect **YZ** at **Q**.
Use (1) to construct a square of area equal to a rectangle whose sides are equal to **YQ** and **XK**.
This is the square required.

Proof. Area of polygon = area of △**XYZ**.
$= \tfrac{1}{2} YZ \cdot XK$.
$= YQ \cdot XK$ = square.

Q.E.F.

Construction 32

(i) Construct a triangle equal in area to one given triangle and similar to another given triangle.
(ii) Construct a polygon equal in area to one given polygon and similar to another given polygon.

(i) *Given* two △s **ABC, PQR**.
To Construct a △**XBZ** equal to △**ABC** and similar to △**PQR**.
Suppose △**PQR** placed with **QR** parallel to **BC**.

Through **A** draw a line **AD** parallel to **BC**.
Through **B** draw **BH** parallel to **QP** to meet **AD** at **H**.
Through **H** draw **HK** parallel to **PR** to meet **BC** at **K**.

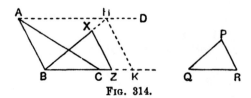

Fig. 314.

Construct the mean proportional **BZ** to **BC**, **BK**.
Through **Z** draw **ZX** parallel to **KH** to meet **BH** at **X**.
Then **XBZ** is the required triangle.

Proof. By parallels, $\triangle$**XBZ** is similar to $\triangle$**HBK** and
∴ to $\triangle$**PQR**.

Also $\dfrac{\triangle \text{XBZ}}{\triangle \text{HBK}} = \dfrac{\text{BZ}^2}{\text{BK}^2} = \dfrac{\text{BC}\cdot\text{BK}}{\text{BK}^2} = \dfrac{\text{BC}}{\text{BK}} = \dfrac{\triangle \text{ABC}}{\triangle \text{HBK}}$.

∴ $\triangle$**XBZ** = $\triangle$**ABC**.

Q.E.F.

(ii) *Given* two polygons **F** and **OSTUV**.
  To Construct a polygon **OS'T'U'V'** similar to **OSTUV** and equal to **F**.

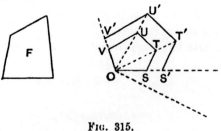

Fig. 315.

Reduce the two polygons **F** and **OSTUV** to equivalent triangles **ABC**, **PQR** respectively and proceed as in (i). [See Fig. 314.]

On **OS** take a point **S'** such that $\dfrac{\text{OS}'}{\text{OS}} = \dfrac{\text{BZ}}{\text{QR}}$.

On **OS'** construct the polygon **OS'T'U'V'** similar to **OSTUV**.
Then **OS'T'U'V'** is the polygon required.

CONSTRUCTIONS FOR BOOK IV   297

Proof. $\dfrac{\triangle OS'T'U'V'}{\triangle OSTUV} = \dfrac{OS'^2}{OS^2} = \dfrac{BZ^2}{QR^2} = \dfrac{\triangle XBZ}{\triangle PQR} = \dfrac{\triangle ABC}{\triangle PQR} = \dfrac{F}{\triangle OSTUV}$

∴ OS'T'U'V' = F.

Q.E.F.

Note the use made of Theorems 58, 59.

CONSTRUCTION 33

Construct a circle to pass through two given points and touch a given line.

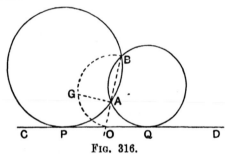

FIG. 316.

Given two points **A, B** and a line **CD**.
To Construct a circle to pass through **A, B** and touch **CD**.
Join **AB** and produce it to meet **CD** at **O**.
Construct the mean proportional **OG** to **OA, OB**, and cut off from **CD** on each side of **O** parts **OP, OQ** equal to **OG**.
Construct the circles through **A, B, P** and **A, B, Q**.
These are the required circles.
Proof. Since $OA \cdot OB = OG^2 = OP^2 = OQ^2$,
OP, OQ are tangents to the circles **ABP, ABQ**.

Q.E.F.

Note that the method fails if **AB** is parallel to **CD**. This special case forms an easy exercise.

CONSTRUCTION 34

Construct a circle to pass through two given points and touch a given circle.

Given two points **A, B** and a circle **S**.

To Construct a circle to pass through **A, B** and touch **S**.

Construct any circle to pass through **A, B** to cut **S** at **C, D** say

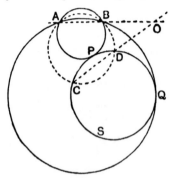

Fig. 317.

Produce **AB, CD** to meet at **O**.
From **O**, draw the tangents **OP, OQ** to **S**.
Construct the circles through **A, B, P** and **A, B, Q**.
These are the required circles.

Proof. **OA . OB = OC . OD**, property of intersecting chords.
$$= OP^2 = OQ^2,$$ tangent property.

∴ **OP, OQ** are tangents to the circles **A, B, P** and **A, B, Q**.
∴ these circles also touch **S**.

Q.E.F.

Construction 35

Construct a circle to pass through a given point and touch two given lines.

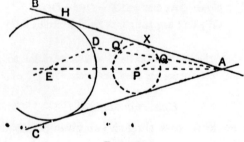

Fig. 318.

CONSTRUCTIONS FOR BOOK IV 299

Given two lines **AB**, **AC** and a point **D**.

To Construct a circle to touch **AB**, **AC** and pass through **D**.

[The centres of all circles touching **AB**, **AC** lie on a bisector of $\angle$ **BAC**.]

Draw any circle touching **AB**, **AC** and let **P** be its centre; **P** being in the same angle **BAC** as **D**.

Join **AD** and let it cut the circle at **Q**, **Q'**.

Draw **DE** parallel to **QP** to meet **AP** at **E**.

With centre **E** and radius **ED**, describe a circle.

This circle will touch **AB**, **AC**.

Proof. If **EH**, **PX** are the perpendiculars from **E**, **P** to **AB**.

$$\frac{EH}{PX} = \frac{EA}{PA} = \frac{ED}{PQ}; \text{ but } PX = PQ.$$

$\therefore$ **EH = ED**.

$\therefore$ circle, centre **E**, radius **ED**, touches **AB** at **H**.

Similarly it may be proved to touch **AC**.

A second circle is obtained by drawing **DE'** parallel to **Q'P** to meet **AP** at **E'**.

Q.E.F.

ANOTHER METHOD.—Take the image of **D** in the bisector of $\angle$ **BAC**, call it **D'**. By the method of Constr. 33, draw a circle to pass through **D**, **D'** and to touch **AB**; this circle will then touch **AC**.

Construction 36

Construct a circle to touch two given lines and a given circle.

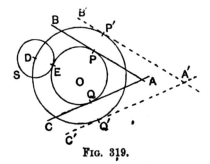

Fig. 319.

Given two lines **AB**, **AC** and a circle **S**, centre **D**, radius r.

To Construct a circle to touch **AB**, **AC**, and **S**.

Draw two lines **A'B'**, **A'C'** parallel to **AB**, **AC** and at a distance *r* from them.

By Constr. 35, draw a circle to touch **A'B'**, **A'C'** and to pass through **D**. Let **O** be its centre.

With **O** as centre, draw a circle to touch **AB**. This circle will also touch **AC** and **S**.

Proof. Let **P'**, **Q'** be the points of contact with **A'B'**, **A'C'**.

Let **OP'**, **OQ'**, **OD** cut **AB**, **AC**, **S** at **P**, **Q**, **E**.

Then $PP' = QQ' = r = ED$; but $OP' = OQ' = OD$.

∴ $OP = OQ = OE$ and **OP**, **OQ** are perp. to **AB**, **AC**.

∴ the circle, centre **O**, radius **OP**, touches **AB**, **AC**, **S**.

Note.—There are in all four solutions: this construction gives two solutions, since two circles can be drawn to touch **A'B'**, **A'C'** and pass through **D**. And by drawing **A'B'**, **A'C'** at distance *r* from **AB**, **AC** on the other side, two other solutions are obtained.

Construction 37

Bisect a triangle by a line parallel to one side.

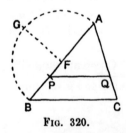

Fig. 320.

Given a triangle **ABC**.

To Construct a line parallel to **BC**, cutting **AB**, **AC** at **P**, **Q** so that **PQ** bisects △**ABC**.

Bisect **AB** at **F**.

Construct the mean proportional **AG** between **AF**, **AB**.

From **AB** cut off **AP** equal to **AG**.

Draw **PQ** parallel to **BC**, cutting **AC** at **Q**.

Then **PQ** is the required line.

Proof. $\dfrac{\triangle APQ}{\triangle ABC} = \dfrac{AP^2}{AB^2} = \dfrac{AF \cdot AB}{AB^2} = \dfrac{AF}{AB} = \dfrac{1}{2}.$

CONSTRUCTION 38

Divide a given line into two parts so that the rectangle contained by the whole and one part is equal to the square on the other part.

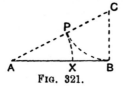
FIG. 321.

Given a line **AB**.

To Construct a point **X** on **AB** so that $AB \cdot BX = AX^2$.
Draw **BC** perpendicular to **AB** and equal to $\tfrac{1}{2}AB$.
Join **CA**.
From **CA** cut off **CP** equal to **CB**.
From **AB** cut off **AX** equal to **AP**.
Then **X** is the required point.

Proof. Let $AB = 2l$.
∴ $BC = l$.
∴ $AC^2 = 4l^2 + l^2 = 5l^2$.
∴ $AC = l\sqrt{5}$; but $CP = CB = l$.
∴ $AP = l(\sqrt{5} - 1)$.
∴ $AX = l(\sqrt{5} - 1)$.
∴ $BX = 2l - l(\sqrt{5} - 1) = l(3 - \sqrt{5})$.
∴ $AB \cdot BX = 2l \cdot l(3 - \sqrt{5}) = l^2(6 - 2\sqrt{5})$
and $AX^2 = l^2(\sqrt{5} - 1)^2 = l^2(6 - 2\sqrt{5})$,
∴ $AB \cdot BX = AX^2$.

CONSTRUCTION 39

Construct an isosceles triangle, given one side and such that each base angle is double of the vertical angle.

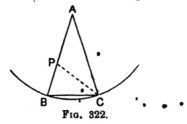
FIG. 322.

Given a side **AB**.

To Construct a triangle **ABC** such that **AB = AC** and ∠ **ABC** = ∠ **ACB** = 2 ∠ **BAC**.

With centre **A** and radius **AB** describe a circle.
On **AB** construct a point **P** such that **AB . BP = AP²**.
Place a chord **BC** in the circle such that **BC = AP**.
Join **AC**.
Then **ABC** is the required triangle

Proof. **AB . BP = AP²**, but **AP = BC**.
∴ **AB . BP = BC²**.
∴ **BC** touches the circle **APC**.
∴ ∠ **BCP** = ∠ **CAP**.
∴ △s **BCP, BAC** are equiangular [∠ **ABC** is common]
But **AB = AC**,   ∴ **CB = CP**.
But **CB = AP**,   ∴ **CP = PA**.
∴ ∠ **PAC** = ∠ **PCA**.
But ∠ **PAC** = ∠ **PCB**, ∴ ∠ **BCA** = 2 ∠ **PAC** or 2 ∠ **BAC**.
∴ ∠ **ABC** = ∠ **BCA** = 2 ∠ **BAC**.

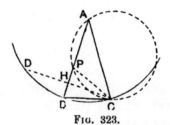

Fig. 323.

Note.—Since the angles of a triangle add up to 180°.
∠ **ABC** = ∠ **BCA** = 72° and ∠ **BAC** = 36°.
∴ **BC** is the side of a regular decagon inscribed in the circle.

From **C**, draw **CH** perpendicular to **AB** and produce it to meet the circle at **D**; then **CH = HD** and ∠ **CAD** = 72°.
∴ **CD** is the side of a regular pentagon inscribed in the circle.

The following result is useful :—
If *p* and *d* are the lengths of the sides of a regular pentagon

CONSTRUCTIONS FOR BOOK IV   303

and a regular decagon inscribed in a circle of radius a, then $p^2 = a^2 + d^2$.

In Fig. 323, let $AB = a$, $CD = p$, $CB = d$; it is required to prove that $p^2 = a^2 + d^2$.

Since $AB \cdot BP = BC^2$ and $BP = BA - AP = BA - BC = a - d$.

∴ $a(a - d) = d^2$ or $a^2 - ad - d^2 = o$.

From $\triangle CHB$, $CH^2 + HB^2 = CB^2$; but $CH = \tfrac{1}{2}CD = \tfrac{1}{2}p$ and $HB = \tfrac{1}{2}PB = \tfrac{1}{2}(a - d)$.

∴ $\tfrac{1}{4}p^2 + \tfrac{1}{4}(a - d)^2 = d^2$.
∴ $p^2 + a^2 - 2ad + d^2 = 4d^2$.
∴ $p^2 = 3d^2 + 2ad - a^2$.
∴ $p^2 = a^2 + d^2 - 2(a^2 - ad - d^2)$.
∴ $p^2 = a^2 + d^2$, since $a^2 - ad - d^2 = o$.

CONSTRUCTION 40

Inscribe (i) a regular pentagon; (ii) a regular decagon in a given circle.

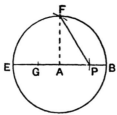

FIG. 324.

Let A be the centre and EAB a diameter of the given circle.

Let AF be a radius perpendicular to AB.

Bisect AE at G.

With G as centre and GF as radius, describe a circle, cutting AB at P; join PF.

Then AP and PF are equal in length to the sides of a regular decagon and a regular pentagon inscribed in the circle.

The regular figures are therefore constructed by placing chords in the circle end to end equal to these lines.

Proof. From GF cut off GR equal to GA; from FA cut off FS equal to FR.

Then by Constr. 38, $FA \cdot AS = FS^2$.
Now $GR = GA$ and $GP = GF$, $\therefore$ $AP = RF = SF$.
But $AF = AB$, $\therefore$ $BP = AS$.
$\therefore$ $BA \cdot BP = AP^2$.
$\therefore$ by Constr. 39, AP is equal to a side of the regular decagon.
But $AP^2 + AF^2 = PF^2$.
$\therefore$ PF is equal to a side of the regular pentagon. (See pp. 302, 303.)

NOTES

GLOSSARY AND INDEX

ACUTE angle: any angle less than 90°.
Alternate angle, 5.
Altitude: the altitude of a triangle is the perpendicular from any vertex to the opposite side.
Angle in a semicircle: an angle whose vertex lies on the circumference and whose arms pass through the extremities of a diameter.
Apollonius' theorem, 226.
Arc of a circle: any part of the circumference.
Area of circle, 86.
Area of triangle and trapezium, 27.

Bisect: divide into two equal parts.

Centroid, 98.
Chord: the line joining any two points on the circumference of a circle.
Circle: the locus of a point which is at a constant distance (called the *radius*) from a fixed point (called the *centre*) is called the *circumference* of a circle.
Circumcentre, 97.
Common tangents, 283
Complementary angles: angles whose sum is 90°.
Concentric: having the same centre.
Congruent: equal in all respects. The symbol is ≡.
Corresponding angles, 5.
Cyclic quadrilateral: a quadrilateral whose four corners lie on a circle.

Decagon: a figure with ten sides.
Degree: $\frac{1}{90}$th part of a right angle.
Depression, angle of, 145.
Diagonal: the line joining two opposite corners of a quadrilateral.

Diameter: a chord of a circle passing through the centre.

Elevation, angle of, 145.
Equilateral: having all its sides equal.
Equivalent: equal in area.
Excentre, 97.
External bisector: if **BAC** is an angle and if **BA** is produced to **X**, the line bisecting ∠ **CAX** is called the external bisector of ∠ **BAC**.

Hexagon: a figure with six sides.
Horizontal line: a line perpendicular to a vertical line.
Hypotenuse: the side of a right-angled triangle opposite the right angle.

Identities, geometrical, 228.
Image, 93.
Incentre, 97.
Isosceles triangle: a triangle with two sides equal.

Locus, 248.

Mean proportional, 121.
Median: the line joining a vertex of a triangle to the mid-point of the opposite side.
Mensuration formulæ, 86.

Nine point circle, 102.

Obtuse angle: an angle greater than 90° and less than 180°.
Octagon: a figure with eight sides.
Orthocentre, 98.

Parallel lines, 208.
Parallelogram, 22.

Pedal triangle, 98.
Pentagon : a figure with five sides.
Perimeter : the sum of the lengths of the sides bounding a figure.
Perpendicular : at right angles to.
Playfair's axiom, 208.
Projection, 224.
Proportional (third or fourth), 290.
Pythagoras' theorem, 222.

Rectangle, 22.
Reflection, 93.
Reflex angle. an angle greater than 180°.
Regular polygon : a polygon having all its sides and all its angles equal.
Rhombus, 22.
Right angle, 205.

Sector of a circle : the area bounded by two radii of a circle and the arc they cut off.
Segment of a circle : the area bounded by a chord of a circle and the arc it cuts off ; a segment greater than a semicircle is called a *major segment*, if less a *minor segment*.
Similar, 257.
Square, 22.
Supplementary angles : angles whose sum is 180°.
Symbols : $=$ equal in area.
 $\equiv$ congruent.
 $\sim$ the difference between X and Y is represented by X $\sim$ Y.
 $>$ greater than.
 $<$ less than.
 $\angle$ angle.
 $\triangle$ triangle.
 $\parallel^{\text{gram}}$ parallelogram.
 $\bigcirc^{\text{ce}}$ circumference.

Tangent, 243.
Trapezium, 22.

Vertical line : a line which when produced passes through the centre of the earth.

ANSWERS.

1. Where only one form of unit occurs in the question, the nature of the unit is omitted in the Answer.
2. Answers are not given when intermediate work is unnecessary.
3. Results are usually given correct to three figures, and for angles to the nearest quarter of a degree.

EXERCISE I (p. 2)

5. 6; 11; 22.　　7. 135°.　　8. 83°; $112\frac{1}{2}$°; 167°.
9. (iv) 300°; (v) 990°.　　10. 20°.　　12. (ii) 65°.
13. 120°.　　15. 120°.　　16. 72°.　　17. 72°.
18. 120°.　　19. $247\frac{1}{2}$°.　　20. 5°.　　24. 40.
25. 110°; $149\frac{1}{2}$°.　　26. 15°.　　27. 46°.　　28. 111°.
29. $111\frac{1}{2}$°.　　30. 251°.　　31. $180 - x$.　　32. $90 + \frac{1}{2}x$.

EXERCISE III (p. 10)

5. 122°.　　6. 93°.　　7. 80°.　　10. 36.　　12. 80°.
13. 80.　　14. Least is 36°.　　15. 8°.　　16. 37°.　　17. 86°.
19. $2x - 180$°.　　20. 120°.　　21. $\frac{1}{2}(x - y) + 90$°.　　25. 162°.
27. $y = \dfrac{6x}{8 - x}$, $y = 6, 10, 18, 42$.　　28. 6.　　31. $x = c - a - b$.
32. $x = b - a - c$.　　33. $x = a + b + c$.

EXERCISE IV (p. 16)

3. (i) 90°, 45°; (ii) 72°, 36°.　　5. 50, 60, 70°.　　6. $x = 360 - 2y$.
7. $x = 60 \pm \frac{1}{2}y$.　　9. 36°.　　33. $25\frac{5}{7}$°.

EXERCISE V (p. 23)

5. 68°.　　7. 62°.　　23. $67\frac{1}{2}$°.

EXERCISE VI (p. 28)

1. 7·5.   2. 17·5.   3. 4·8.   4. 4.   5. 42.   6. 44.
9. 4·8.   10. 12.   11. 6·75.   12. 10·5.   13. 3·75.
14. 4·5; 4.   15. 4·8.   16. 15.   17. 4·8; 4·8.   18. 4·4.   19. 26.
20. 8.   21. 6·2; 20.   22. 4′ to mile; $\frac{2}{3}''$.
23. $\frac{1}{2}(xq + xr + yp + yq)$.   24. $\frac{1}{2}(pr + qr + qs)$.   25. $\frac{pq}{r}$.
26. 24; 12; 36.   28. $\frac{1}{2}(xy - ef)$.   29. 5; 10.
30. (i) 4; (ii) 5; (iii) 5·5; (iv) $\frac{1}{2}ac$; (v) $\frac{1}{2}(ad - bc)$.   31. (i) 10; (ii) 11.
32. (i) 3·3; (ii) 6·4.   33. (i) 14·7, 5·88; (ii) 57·2, 14·3.   34. 5·56.

EXERCISE VII (p. 38)

1. 13.   2. 8.   3. 5·66.   4. 32·25.   5. $9\frac{1}{3}$.   6. 5·83.
7. 217.   8. 4·77.   10. 30.   11. 14970.   12. 17·3; 1·975 ft.
13. 21·1.   14. 16·2 mi.   15. 60 yd.   16. 4·47.   17. 5.   18. 5.
19. 6·93.   20. 2·89.   21. 5.   22. 5; 7.   23. 13.   26. $6\frac{2}{3}$.
27. $8\frac{3}{8}$.   28. 55·2.   29. 5·46.   30. 3·57.   31. 7.
33. 9·16.   34. 8·66.   35. 26·8.   36. 18.   37. 6·24.
39. Each side 60 sq. in.; 11·7 in.   40. 7·34.

EXERCISE VIII (p. 44)

1. (i), (ii), (iv).   2. 19.   3. $1\frac{2}{3}$; 2·67.   4. 5·85; 6·84.
5. 11; 1; 6·93.   6. 42·43.   7. 6·63.   8. 12·2.
10. Yes.   13. 3·5.   14. 5·45; 6·52; 7·97.
15. 9·17.   16. 10.   17. 12·7.

EXERCISE X (p. 49)

13. 7.

EXERCISE XI (p. 52)

21. 12″; 17″.

EXERCISE XII (p. 57)

1. 9·16.   2. 13.   3. 11·5.   4. $7\frac{1}{24}$.   5. 8·58, 0·58.
6. 5·38.   7. 3·46.   8. 5.   9. 4.   10. 8.
11. 4·8.   12. 3·12.   13. $x^2 + xy = a^2 - b^2$.   14. 5·22.
15. $\dfrac{x^2 + y^2 + z^2 - 2xz}{2(x - z)}$.

EXERCISE XIII (p. 62)

1. 40°.   2. 55°.   3. 110°.   4. 37°.   5. 107°.   6. 100°; 110°.
7. 54°; 99°.   8. 105°.   9. 72°.   10. 124°.   11. 54°.   12. 105°.

EXERCISE XIV (p. 68)

1. 62°.   2. 117°.   3. 26°, 8°.   4. 58°, 64°.   5. 103°, 90°, 77°, 90°.
6. 94°, 8°.   7. 120°.

EXERCISE XV (p. 72)

1. 30°, 45°, 105° or 15°, 30°, 135°.   2. $7\frac{1}{2}$°, $22\frac{1}{2}$°, 150° or $22\frac{1}{2}$°, 30°, $127\frac{1}{2}$°.
4. 3 : 1.   5. 46°, 37°.

EXERCISE XVI (p. 77)

1. 3.   2. 2·5, 1·5, 4·5.   3. 8, 4, 3.   4. 5·3, 3·6, 4·5.
5. 10·5, 1·5.   6. 6   7. $1\frac{7}{8}$.   8. 32, 8.
9. 3.   10. 1·5, 2·5.   11. ·5, 2·5.   12. 12.
13. 19·1, 12.   14. 7, 1.   15. 1·45, 11·125.   16. $5 - 3\sqrt{2} = 0.757'$.
18. 1·44, 36.   19. $2\frac{1}{8}$.   20. $1+\sqrt{2}=2.41$.

EXERCISE XVIII (p. 87)

1. 25·1 in., 50·3 sq. in. ; 628 yd., 31,420 sq. yd.   2. 0·8.   3. 1·1.
4. 2·1.   5. 5·89.   6. 4·57.   7. 57° 18′.   8. 3·2.
9. 158·5.   11. 84·8.   12. 21·5.   13. 628 ; 408.   14. $3\frac{3}{8}$.
15. 25.   16. 314 ; 204.   17. $\frac{3}{8}$.   18. 288°.   19. 48 ; 96
20. 65·4 ; 78·5.   21. 100,000,000 sq. m. ; $\frac{1}{2}$.   22. 8·2.   23. 9·21.
24. 20·1.   25. $2\frac{3}{4}$.   26. 78·5.   27. 514 ; 500 ; 9·0.
28. 119 ; 44·0.   29. 77·4.   30. 828·5 sq. ft.   34. 11·8.
35. 29·3.   36. 102·5.   37. 8 ; 14 ; $1\frac{5}{7}$.   38. 6·86 ; 137 ; 186.

EXERCISE XIX (p. 94)

32. 20 in.

EXERCISE XXI (p. 106)

2. (iv) $1\frac{1}{4}$.   5. $1\frac{1}{4}$; 0 or 1.   7. 6.   8. $\frac{1}{16}$.
10. 3·2.   11. 6.   15. 2 : 5 ; 1 : 2.   16. 1·6″.   18. 3·2.
21. $\frac{x \sim y}{2(x+y)}$.   22. $\frac{2cy}{x^2-y^2}$; $\frac{x-y}{x+y}$.   23. $AF = \frac{x(p+q+r+s)}{q+r}$.   25. $\frac{1}{\lambda-1}$.
27. $4\frac{1}{8}$.   28. 1·6.   29. $\frac{a\mu+b\lambda}{\lambda+\mu}$.   41. 1.

EXERCISE XXII (p. 112)

1. 120.   2. 4 ft.   4. $10^5 \times 8.6$ mi. ; $10^5 \times 2.3$ mi.   5. 6′ 8″.
6. 66.   7. 14·4″.   8. 6·4, 7·2 cms.   9. $22\frac{1}{2}$.   10. 1·5, $3\frac{3}{4}$.
11. 5.   12. $8\frac{1}{4}$.   13. (i) $\frac{3}{4}, \frac{7}{8}$; (ii) $6\frac{1}{2}$; (iii) $2\frac{3}{4}, 1\frac{1}{2}$; (iv) $5\frac{1}{4}$; $3\frac{3}{8}$.
15. 2·4.   16. 18, 8.   17. 7·2.   18. 14.   19. $3\frac{1}{2}$, 11.
20. 12·8, 5.   21. $8\frac{3}{4}$.   22. 4.   23. (i) $2\frac{1}{2}$; (ii) $7x+5y=35$.
24. 2·9.   25. 12.   26. $1\frac{3}{4}$.   27. 6, 11.   28. $3\frac{1}{8}$.
29. (i) 54′, 24′ ; (ii) 13″.   30. $3\frac{4}{11}$.   31. $y = \frac{fx}{u-f}$.   32. $y = \frac{fx}{u-f}$.
34. $y = \frac{fx}{u-f}$.   35. $13\frac{1}{8}$.

EXERCISE XXIII (p. 122)

2. 6. **3.** $\frac{7}{8}$. **4.** 10. **5.** 2 or 10.
6. (i) 6 ; (ii) 12 ; (iii) 2·31 ; (iv) $21\frac{9}{11}$. **7.** $4\frac{1}{2}$, $6\frac{1}{4}$. **8.** $4\frac{1}{8}$.
11. 7·07 ; 13·04. **12.** 0·707. **13.** $\dfrac{p^2 r}{q^2 - p^2}$, $\dfrac{pqr}{q^2 - p^2}$.

EXERCISE XXIV (p. 127)

1. 12 sq. ft. **2.** 40. **5.** 9. **6.** $101\frac{1}{4}$. **8.** 4·2.
9. 3·75 sq. in. **10.** 16 : 4 : 3 : 9. **12.** £$5\frac{1}{4}$. **13.** $4\frac{1}{7}$. **15.** 512.
16. 1·024. **17.** 6. **18.** 2s. 3d. **19.** $9\frac{1}{7}$. **21.** 40·5 ; 162.

EXERCISE XXV (p. 132)

1. 3, 15. **2.** 3·35. **4.** 12. **5.** $9\frac{3}{4}$. **6.** 3 sq. in.

EXERCISE XXVI (p. 135)

30. 4·8. **59.** 81° 45' or 14° 40'.

EXERCISE XXVII (p. 145)

1. 94·3. **2.** 7140. **3.** 13' 9". **4.** 10' 8". **5.** 32.
6. 2·77. **7.** S. 37° W. ; 5·17 mi. **8.** 7·0 mi. N. 34° W.
9. 8·42 mi. N. 12° W. **10.** E. $36\frac{3}{4}$° N. **11.** 34·8 mi. N. $31\frac{1}{2}$° W.
12. 10·5. **13.** 321. **14.** 91·9. **15.** 85·3. **16.** 2·59. **17.** 84·0.
18. 326. **19.** 31°. **20.** E. 59° S. **21.** 177. **22.** 137. **23.** 34·4.

EXERCISE XXVIII (p. 148)

3. 3·36. **6.** 2·5. **7.** 6·13. **8.** 2·83.

EXERCISE XXIX (p. 150)

1. (i) 36° 50' ; (iii) 2·59 ; (iv) 2·93 ; (v) 4·79 ; (vii) 6·68 ; (viii) 5·66, 3·53 ; (xii) 11·3 ; (xiii) 8·49 ; (xiv) 8·87 ; (xv) $104\frac{1}{4}$°. **8.** 5·74. **9.** 5·23.
10. $106\frac{1}{4}$°. **11.** $49\frac{1}{2}$°. **12.** $62\frac{1}{4}$°. **13.** 5·41. **14.** 2·55. **15.** 7·13, 3·63.
16. $49\frac{1}{2}$°. **17.** (i) 4·96 ; (ii) 6·76 ; (iii) 5·18 ; (iv) $63\frac{1}{4}$° ; (v) 3·82.
18. (i) $25\frac{1}{4}$° ; (ii) 8·25 ; (iii) 6 ; (iv) 6·21. **19.** 8·64. **20.** 3·53.
21. 4·67. **22.** 7. **23.** 6·09. **24.** 6·16. **25.** 4·26. **26.** 4·96.
27. 4·62. **28.** (i) 7·67 ; (ii) 7·10 ; (iii) 10·1 ; (iv) 4·78 ; (v) 7·82 ; (vi) 8·71.
(vii) 6·64. **29.** 6·22. **30.** 5·34.

EXERCISE XXX (p. 153)

10. 1·63. **11.** $21\frac{3}{4}$°.

EXERCISE XXXI (p. 155)

1. (i) 10 ; (ii) 50·0 ; (iii) 14·7 ; (iv) 6 ; (v) 48 ; (vi) 9·43 ; (vii) 45·1 ; (viii) 28 ; (ix) 21 ; (x) 18. **2.** 15·0. **3.** 5·75. **4.** 4·57. **5.** 30°.
6. 2·64. **7.** $36\frac{3}{4}$°. **8.** 40°. **9.** 4·07. **10.** 5·80 or 10·6. **13.** 29·1

ANSWERS

EXERCISE XXXII (p. 158)
6. 1·93.   **7.** 3·61.   **8.** 6·82".   **9.** $3\tfrac{1}{880}$.   **10.** 3·17.

EXERCISE XXXIII (p. 161)
5. 6·65.   **17.** 0·64, 1·16, 1·93, 5·80.   **18.** 1·46.   **24.** 2·13.   **26.** 3·11.
27. 1·94.   **28.** 4·61.

EXERCISE XXXIV (p. 172)
6. 4·47.   **9.** 3·20.   **14.** 2·66.   **15.** 1·56.   **16.** 5·80.   **17.** 1·32.
18. 8·13.   **24.** 5·60, 2·14.   **25.** 6·06, 4·02.   **29.** 5·87, 2·23.
30. 6·89, 4·89.   **35.** 4·16.   **37.** $11\tfrac{3}{4}°$.

EXERCISE XXXV (p. 175)
1. 7·5.   **2.** 7·2.   **7.** (i) 2·89; (ii) 10·3.   **11.** 4·12; 1·21.   **16.** 3·63.
20. 2·27.   **21.** 4·55.   **22.** 2·68.   **23.** 5·36.

EXERCISE XXXVI (p. 178)
1. 6·325.   **3.** 6·08.   **4.** 7·36 or −1·36.   **5.** 5·29.   **6.** 3·29.
7. 5·00.   **19.** $x = 7\cdot22$ or $-2\cdot22$, $y = 2\cdot22$ or $-7\cdot22$.

EXERCISE XXXVII (p. 180)
3. 10.   **4.** 5·78.   **5.** 4·81.   **7.** 3·83.

REVISION PAPERS (p. 181)
1. 300°.   **6.** 112°.   **10.** 110°.   **29.** 75°.   **33.** $x = 540 - a - b - c$.
37. $67\tfrac{1}{2}°$.   **41.** $z = 180 - a - b - x - y$.   **42.** $\dfrac{12}{n}$ rt. angles.   **46.** 80°.
49. 3·75.   **53.** $\tfrac{1}{2}(xy+yz)$.   **56.** 4·24.   **57.** 13·4 (5).   **61.** $2\tfrac{3}{8}$.
64. 5·5, 2·5, 17·3.   **65.** $\tfrac{1}{2}[p(y+r)+q(r+s)+x(s-y)]$.
68. 13".   **69.** 15, 9.   **72.** 9.   **73.** 300.   **77.** 7·5.   **78.** 2·16.
81. 12; 5·66.   **85.** 2.   **88.** (ii) $\sqrt{x^2 - 8x + 416}$; (iii) $x > 6\tfrac{1}{2}$.
89. 2.   **99.** 47°.   **102.** 60°, 80°.   **106.** $\dfrac{a^2 + 4h^2}{4h}$.
107. 55°, 40°.   **109.** 13.   **115.** 15°.   **118.** E. 25° N.
131. 17.   **133.** on AB 10, on CD 20.   **136.** 43·2.   **137.** 5.   **144.** $\dfrac{2V}{S}$.
147. $\tfrac{3}{4}$.   **155.** $9\tfrac{1}{2}, \tfrac{1}{2}$.   **159.** $6\tfrac{3}{4}$ in.   **167.** 6, 10, 14 in.
170. 132.   **174.** 4·47.   **175.** 24·4 in.   **177.** 6, 5.
179. 2·2.   **182.** 0·69 or 23·3.   **183.** $5\tfrac{1}{4}, 2\tfrac{1}{11}$.   **187.** 2.
189. 3·2, 1·2, 4·4.   **194.** 68·7.   **195.** 2, $2\tfrac{3}{4}$.
197. 4.   **199.** 4·8.

PRINTED BY
MORRISON AND GIBB LIMITED
EDINBURGH

Lightning Source UK Ltd.
Milton Keynes UK
02 December 2009

147053UK00001BA/91/A